From Artifact to Habitat

Research in Technology Studies

From Artifact to Habitat

Studies in the Critical Engagement of Technology

Research in Technology Studies,
Volume 3

EDITED BY
Gayle L. Ormiston

Bethlehem: Lehigh University Press
London and Toronto: Associated University Presses

© 1990 by Associated University Presses, Inc.

Associated University Presses
440 Forsgate Drive
Cranbury, NJ 08512

Associated University Presses
25 Sicilian Avenue
London WC1A 2QH, England

Associated University Presses
P.O. Box 488, Port Credit
Mississauga, Ontario
Canada L5G 4M2

The paper used in this publication meets the requirements
of the American National Standard for Permanence of Paper
for Printed Library Materials Z39.48-1984.

Library of Congress Cataloging-in-Publication Data

From artifact to habitat : studies in the critical engagement of
technology / edited by Gayle L. Ormiston.
     p.   cm.—(Research in technology studies ; v. 3)
   Essays from a conference on The Mediation of technology, held at
the University of Colorado, Colorado Springs, 15–16 Nov. 1985.
   Includes bibliographical references.
   ISBN 0-934223-09-2 (alk. paper)
   1. Technology—Philosophy—Congresses.   I. Ormiston, Gayle L.,
1951– . II. Series.
T14.F76   1990
601—dc20   88-46174
                                                           CIP

PRINTED IN THE UNITED STATES OF AMERICA

For Lynn and Rachel Alison

Contents

Foreword
  STEPHEN H. CUTCLIFFE and STEVEN L. GOLDMAN   9

Acknowledgments   11

Introduction
  GAYLE L. ORMISTON   13

List of Contributors   25

Three Ways of Being-With Technology
  CARL MITCHAM   31

Technological Consciousness and the Modern
Understanding of the Good Life
  JAMES L. WISER   60

Technology and the Denial of Mystery: The Sacrilization of
the Familiar
  DAVID LOVEKIN   74

Issue and Presentation: Technology and the Creation of
Concepts
  GAYLE L. ORMISTON   102

The Autonomy of Technology
  JOSEPH C. PITT   117

The Labor-Saving Device: Evidence of Responsibility?
  EDMUND F. BYRNE   132

The Alienation of Common Praxis: Sartre's *Critique* and the
Weberian Theory of Bureaucracy
  FREDERIC L. BENDER   155

Contents

Artificial Representationalism: Necessary(?) Constraints on
Computer Models of Natural Language Understanding
  PHILIP A. GLOTZBACH   175

From Artificial to Natural Intelligence: A Philosophical
Critique
  LAWRENCE DAVIS   196

Select Bibliography   209

Index   215

Foreword

STEPHEN H. CUTCLIFFE
STEVEN L. GOLDMAN

This is the third volume in the series *Research in Technology Studies* published by Lehigh University Press. Each volume consists of a group of invited essays on a theme set by a guest editor. For this volume, Gayle L. Ormiston has invited writers of diverse interests to address the philosophy of technology, a subject only now beginning to receive serious attention in the United States. The richness of their response is suggested by an inventory of the thinkers whose ideas are examined in the contributed essays: the Xenophonic and the Platonic Socrates, Plato, and Aristotle; Bacon, Descartes, Galileo, and Vico; Rousseau, Wordsworth, Poe, and Stapledon; Ure, Weber, Mayo, and Jung; Nietzsche, Heidegger, Sartre, Ellul, and Derrida; Kuhn, Braverman, Katz, and Fodor; Winograd, Schrank, and Gibson. The essays range from attempts at wresting new insights into technology from familiar figures to attempts at exposing the "essence" of technology by reaching out to unexpected figures. They also range from traditional philosophy through literary and linguistic criticism to artificial intelligence research programs. The coherence of the essays lies in the attention each in its own way draws to the profound implications of technology for the "human-all-too-human."

Acknowledgments

The essays in this volume arise from a conference on *The Mediation of Technology*, held at the the University of Colorado at Colorado Springs, 15–16 November 1985. Without the university's support, the conference and, as a consequence, this volume would not have been possible. I thank each of the authors for their contributions to the collection and for their support, patience, and advice during the long period it took to bring the volume into final form. Edmund F. Byrne and Joseph C. Pitt gave permission to reproduce their essays which were orginally published in the *Philosophy and Technology* series edited by Paul Durbin, published by D. Reidel; Paul Durbin also granted his permission to reprint the essay by Pitt. Stephen H. Cutcliffe and Steven L. Goldman provided their valuable suggestions regarding revisions for the final manuscript.

For her advice during the planning stages of the conference, for her unstinting support of the projects I create for myself but which bear directly on her life as well, and for her incisive and helpful criticisms of these projects, I wish to express my gratitude to and deepest appreciation for Lynn Ormiston.

Introduction

GAYLE L. ORMISTON

The pretext for the papers incorporated in this volume is the ubiquity and mediacy of technology: technology appears simultaneously as an artifact of human existence *and* as the medium by which the habitat of human existence comes to be what it is. The alliance of artifice, artifact, and habitat is so fused with (and in) our daily lives that the identity of "technology"—a response to the question "What is technology?"—seems patently obvious: simple and familiar. Why? Because we are immersed in a world composed of and created by technological artifice. We work with and engage particular understandings of technology—pedestrian or otherwise— every day. In doing so, we assume technology's uniformity and comprehensibility. In general, we presuppose an identity of technology and a knowledge of its effects: we act as if technology can be identified as such, as if the concept of "technology" can be articulated apart from our engagement of and with it, apart from our descriptions of it. But, as the notion of a *pretext* suggests—where *praetextus* indicates "outward show or pretense," and *praetexere* signifies "to weave in front, to cloak, disguise, or pretend"—the ubiquity and mediacy of technology do not permit such a facile understanding.

The artifacts of technological innovation—the products of technological artifice and the various grades of clarity that guide our understanding of these techniques, practices, and their effects— already focus our perceptions and orient our comprehensions of the "world" and our "selves"; they focus our understanding of the interaction between technology, so-called, and human existence. The interweaving of artifice and artifact, then, indicate an engagement (or engagement*s*) with technology traversing many different levels of understanding, fashioned prior to any reflective involvement with or critical discourse on technology. In other words, we have already framed and put into play an understanding, a "concept" of technology, especially in terms of its mediacy, by which we translate the

13

discourse *on*, *of*, and *by* the individual or the "subject" according to how it is interlaced already with the discourse on *techné*, or art, and creative production (scientific or other). Under these conditions, technology withdraws into our habitat—into the structures of thought, action, and discourse.

The sensible presence of machinery, gadgets, and techniques designed for the physical manipulation of our environment, along with the effects of these techniques, fixes our attention on certain aspects of what goes by the name "technology." The representation of technology in artifacts and artifice—artifice that traverses computer hardware and software, electronic sensors in our automobiles that indicate an open door, oil pressure, and hazardous road conditions or that coordinate no-lock brake systems, to the automation of routine procedures and deployments of robotics in the home or industry, to the digital quartz watch I wear that is capable of arousing me from a sleep with its alarm, recording how many hours I spend running, and informing me when I am late for a class—conceals and disguises an economy that makes possible the withdrawal and transformation of technology into a habitat.

To advance a claim about the ubiquity and mediacy of technology is to claim that technology cannot be comprehended merely in terms of its "presence" in sensible representations. Instead, technology's ubiquity and mediacy can be traced only in terms of its dissemination, in terms of its already having withdrawn from its objective representations and having become the binding element of the human habitat.

Thus, the question of technology is pressing for two reasons. First, one might expect that we *should* be able to offer an exacting definition of "technology"—to identify its governing scientific, ethical, social and political, and linguistic principles in order to provide an account of its conceptual foundations. But it is for this very reason, because of our expectations, our familiarity with, and our mediate relation to technology in its various forms and manifestations, that the question needs to be asked—always, again. Ludwig Wittgenstein remarks in the *Philosophical Investigations*:

> The aspects of things that are most important to us are hidden because of their simplicity and familiarity. (One is unable to notice something— because it is always before one's eyes.) The real foundations of his inquiry do not strike him at all. Unless *that* fact has at some time struck him.[1]

The second reason is linked intimately to the first. If we are to understand technology, and if we are to comprehend the challenges and questions it poses to individual existence as well as to other social and political structures, then our task becomes one of assembling reminders about the constitution of our habitat, reminders about the community and our engagements in the community. Perhaps, as Joseph Agassi claims, it becomes our task to recognize technology as "any means of social control (of anything)."[2] It is against this backdrop, and out of or *from* our engagement in the community and our comprehension of this engagement, that technology is interpreted, understood, and determined; we appropriate technology, incorporate its techniques into our lives, and in doing so we create ourselves. The ubiquity and mediacy of technology challenges us in just this way: we need to remind ourselves of the principles and values on which we base the organization and orientation of the community—at least with regard to how we participate in it—and technology's role in that organization and orientation, so as not to lose sight of them, or to take them for granted and *obvious*, not to allow them to get in the way of recognition (in the Latin, *ob viam* indicates how that which is exposed, open, or met with plainly stands in the way of or against itself).

What is at stake in the current analysis of technology is not merely the assumed moral values and principles that inform our attempts to control our environment and, as such, technology. What is at stake are the various images we carry forth of our selves and others, and how we translate these images in terms of how we desire to fix the future, to fabricate certain fables about our selves, and that which is unknown to us, in order to legitimate and to provide a foundation for our actions today.

The essays collected in *From Artifact to Habitat: Studies in the Critical Engagement of Technology* embrace the challenge and the question of technology. This collection constitutes a study of the interposition and mediacy of technology and its apposite concepts, as well as a study of the meanings assigned to "technology" and the metaphors used in the analysis of technological concepts. Moreover, the essays included in this volume address different aspects of technology's ubiquity and mediation.

Technology has been broadly construed as: (1) the application of ideas in a practical context for problem solving; (2) a set of organs—that is, a totality of means, gadgets, and physical devices of technical performance; (3) routine procedures followed to accomplish specific tasks at a given level of competence or experise; and (4) the term, concept, or the equipment fashioned in a particular context for a specific purpose, the manipulation of the physical world. However it is construed, presented, or deliberated, technology today reflects the dreams, hopes, and fears and the fluctuating and variegated needs, objectives, and capabilities of those interested in it and those affected by it. The goal of these essays is to establish an interpretive framework by which the questions, issues, themes, and perspectives that constitute the critical study and development of technology can be examined.

Any attempt to provide a framework for interpreting the ubiquity and mediacy of technology involves certain risks. On the one hand, such an attempt must recognize the need to offer a systematic discussion of the current philosophical, political, psychological, literary and artistic, theological and scientific, and linguistic parameters involved in creating the context for any investigation into the fundamental ethical and economic concerns related to the conduct of and application of science and technology. The temptation here is to assume a coherence and continuity in the field of technology studies—even a certain *definition* of "technology," one that governs the identification of a certain problematic, whatever it may be. In other words, it is easy to assume that there is an already fixed set of problems and values, along with strategies for analyzing these issues, that constitutes the subject matter for the philosophy of technology in particular, and for studies in science and technology in general, because a fixed conception of technology is enacted. As the literature and discourse in this field indicate, the issue of technology is an extremely rich and complex one. By focusing on a particular thematic unity found in the contemporary discourse on technology, it is possible to obscure not only the different values and ideas that inform the various efforts to create a field of technology studies, but also the differences among the issues that are identified as pertinent to this field of questioning.

On the other hand, the desire to provide an interpretive framework for comprehending the ubiquity and mediacy of technology forces a shift in the discourse on technology. The attempt to frame the ubiquity of technology is not simply an attempt to recognize this condition; nor is it an attempt to draw attention to technology's ubiquity as fact; nor is it merely a way of accepting technology as a pervasive factor in contemporary life. Instead, to force a shift in thematic emphasis, to refocus the question of technology, is to offer *one other* way (or nine ways, if one takes the essays incorporated into this volume respectively) of rethinking technology, thinking through the concept of technology according to its mediacy, according to the alliance of artifice and artifact and habitat. To force a shift in the discourse is to pursue *one other* way of introducing different strategies for articulating conceptions of technology and for compre-

hending how these concepts mediate, and make possible, our comprehension of the world and the cosmos, others and our*selves*. More
directly, framing technology in terms of its diffuse and pervasive
character is an attempt to rethink the internal and external edifice or
alignment of technological structures and practices.

The essays by Carl Mitcham, James L. Wiser, David Lovekin,
and Gayle L. Ormiston initiate the examination of technology's
structural alignment. Rather than arguing for a position of authority
or dominance in the coupling of humanity and technology, in
"Three Ways of Being-With Technology" Carl Mitcham sketches
three ways in and through which the relationship has developed.
Mitcham develops a historicophilosophical description of three
paths on which humanity has embarked in its attempt to come to
terms with technology: ancient skepticism, the optimism of the
Renaissance and Enlightenment, and the romantic ambiguity or
disquietude with itself and technology. According to Mitcham, the
suspicion with which technology is regarded by some today, and the
accompanying ambiguity toward our ability to monitor and to "be-
with" technology, extends a certain strain of the romantic attitude
that suggests the need for a "new and more comprehensive way of
being-with technology."

James L. Wiser's approach is historical as well. In "Technological
Consciousness and the Modern Understanding of the Good Life,"
Wiser identifies three distinct levels of meaning associated with
"technology." At the outset, Wiser focuses on technological consciousness (level 3), which encompasses the basic assumptions, beliefs, and expectations that give way to and sustain the operation of
technology as "machine" (level 1) and technology as "bureaucracy"
(level 2). Wiser's focus then turns toward a critical analysis of what
is called "modern technology" and the development of a certain attitude, first articulated by Francis Bacon and later by René Descartes,
it engenders regarding humanity's place in the universe. According
to this attitude, the Aristotelian notion of the contemplative life or
the life of reason—*bios theoretikos*—is replaced by technology,
"the life of fabrication," as the "essential means for human self-
actualization," or what would otherwise be known in Aristotelian
terms as the "ultimate standard of human excellence."

Elaborating on an argument presented by Jacques Ellul in *The
Technological Society*, David Lovekin claims that the mentality of *la
technique* or technology binds humanity's imaginative vision to the
here and now. No longer is there any sense of *mystérion* that goes
beyond the grasp of reason and *the* method of science and logic—*la
technique*. For Lovekin, the absence of transcendent mystery can be

linked directly to the emergence of the Cartesian method and mentality, where the objective is to find the "keys" that unlock the mysteries of nature. Even the modern mystery novel, the detective story, and science fiction are seen by Lovekin as manifestations of *la technique* and, as such, cannot be considered "imaginative literature." Drawing on Giambattista Vico's opposition of imagination (*fantasia*) and reason (*ragione*), as well as Ellul's assessment of the technological order, Lovekin claims there is a difference between literature as "technologically familiar" and imaginative literature. Literature that reifies a familiar, "natural," and sacred order, such as Edgar Allan Poe's detective stories and the utopian vision presented in Samuel Butler's *Erewhon*, is unable to provide the culture from which it arises "with a sense of being and purpose, meaningful ritual and a sense of place."[3] What is needed, according to Lovekin, is a literature that transcends rational appearance and presents a new sense of responsibility toward science and technology—a literature that presents a new vision of what can be.

Tracing several etymological and conceptual links between *techné*, art, and translation (metaphor), Gayle L. Ormiston advances the claim that however "technology" is presented or deliberated it is presupposed in and legitimates all questioning, all forms of fabrication. Yet "technology" is not an example of a transcendental category or a fundamental principle or merely one question or problem among others. The word, the network of concepts, the equipment and skills associated with—and that underwrite—the fabrication and use of "technology" designate both a problem and a field of interrogation. Technology is comprehensible, then, only as a relay within a series of relays or an assemblage of concepts that always transforms itself, translating over and over again our concern with the conditions under which we would like to live. In the end, Ormiston asks if this is not "the *issue* of technology—the creation and weaving of intermediary concepts and practices?"

From a perspective granted by our vulgar engagement with technology and the situations that consume our attention, an attitude of "everydayness" as it has been called, most notably in the writings of Edmund Husserl,[4] the artifacts of technological artifice appear "autonomous." Watches and clocks, computers, the furnaces that heat our homes, the generators that provide the electrical current to light our homes and to power our trinkets and toys, and even the running shoes thrown in a corner of a room share the characteristic status of independent things—things that blend into the background of our habitat, always there to mediate our comprehension of that habitat and our dwelling in and with technology. To be sure, technology is sometimes thought of and talked about as a "thing," an agentless force that mediates our lives in ways that can be anticipated, but also in unforeseen ways that prohibit the completion of our projects and the achievement of goals we set for ourselves. Indeed, as Wiser notes in "Technological Consciousness and the Modern Understanding of the Good Life," one way by which technology is interposed in our daily activities most notoriously, as an independent force, is through the organization and administration of "first-level skills, knowledge, and equipment"—or bureaucracy.[5]

The encounter with the creations of technological artifice undeniably takes place in a context configured by a complex network of cultural, linguistic, ethical, social and political, religious, and scientific interests. Moreover, our interpretations of these encounters, our discourses on technology, our attempts to control "technology," unfold within a specific region; they are always local and related to a specific field of possibilities. This is to say, the context or the habitat in which we articulate and act upon our various conceptions of technology is never neutral. Techniques and artifacts are bound intimately, always and already, to a localized organization and administration of skills, knowledge, and equipment. It is in this context that questions arise relating to the autonomy of technology, the responsibility assumed by individuals given the denial of technological determinism, and how technology is absorbed by and into the machinations of social relationships.

In "The Autonomy of Technology," Joseph C. Pitt develops a line of thought contrary to one developed by Jacques Ellul in *The Technological Society*[6] but similar to, although independent of, one argued by Langdon Winner in *Autonomous Technology*[7] and *The Whale and the Reactor*.[8] Pitt argues that to speak of the "autonomy" of technology commits one to a categorial mistake, a fundamentally "wrong way" of discussing and understanding technology and its attendant issues. Technology cannot be an independent "thing" in Pitt's eyes. To accept such a position is to reify technology. Moreover, it perpetuates a view that technology is a "threat" to our lives and the handmaiden of science. In order to grasp just how integral technology is to our everyday lives, our society, and our culture, we need to take into account, and to give an account of, the use of technology, specifically the ends toward which technology is developed and employed. Once our inquiries proceed in this fashion, according to Pitt, we will understand better why there is no such problem as the "autonomy" of technology. Rather, the problem is the use *we* make of "technology."

In "The Labor-Saving Device: Evidence of Responsibility?"

Edmund F. Byrne also addresses the question of the autonomy of technology. Byrne's analysis continues the one inaugurated by Pitt and turns toward the issue of the responsible employment of technology. Byrne claims that if one accepts the view that technology is "runaway" or "autonomous" of human agency, there is theoretically no room for any discussion of moral or even legal responsibility. To be sure, the advance of technology proceeds at such a rate as to supersede and to render obsolete existing techniques. To the "uninvolved observer," then, it *appears* that humans are no longer in control of technology or of their own lives. Byrne's task is to show that even if one concedes certain points in this discussion to "technological determinism," it remains the case that *"some* human beings are clearly responsible agents of technological change."[9] In other words, some individuals "not only consciously seek the development of certain technologies but tolerate or even fully intend anticipated consequences of the introduction of that technology."[10] To demonstrate how this claim can be generalized to apply to any form of technological change, Byrne focuses on "labor-saving devices."

Frederic L. Bender's approach to the relations between technology, autonomy, and responsibility, in "The Alienation of Common *Praxis*: Sartre's *Critique* and the Weberian Theory of Bureaucracy," is framed in terms of technology's relationship to bureaucracy. Bender argues that the critique of mainstream organization theory, despite its many insights, remains inadequate for revealing the reasons why technologized bureaucracy is so fundamentally "against the grain," so basically dehumanizing. He suggests that one might do well to look to the concept of *selfhood* and the "mediated reciprocity" that marks individual existence as it is described in Jean-Paul Sartre's analysis of collectivities, the group-in-fusion, the pledged-group, the organization, and the institution.[11] Bender advances the position that Sartre's analysis offers an alternative, certainly more far reaching in breadth and depth, to contemporary critical theory's approach to the ideology of technology.

Technology establishes the boundaries of our habitat, and in doing so forces us to contend with a complex network of constraints or laws—both physical and metaphysical, or "natural" and "artificial." As we become accustomed to the products of technological artifice, as we recognize that these artifacts constitute the very texture of our existence, some would say we lose sight of the our "natural" habitat, even our "human nature." Technology can be said to isolate us from "nature" through the construction of an "artificial" nature, a world of "autonomous" machines, dehumanizing social relationships that foster self-estrangement, and, even more threatening, "autonomous techniques" that have developed well beyond the workings of the machine.[12] Others would argue that more frequently technology enables us to complete projects and achieve goals that otherwise might be, at the very least, unattainable or unthinkable.[13] Rather than abiding by the opposition of the natural and the artificial, those who understand technology in terms of the alliance between artifice and artifact also understand that the very apparatus of technology disengages such classical dichotomies. Any such confrontation of terms is, at best, fictive and heuristic. Thus, the discourse that revolves around this point of contention, and that hopes in one way or another to resolve the issue, must remain mindful of the condition that makes possible the debate: that condition is the question of technology.

The discourse *about* or *on* artifice and dissimulation, the *logos* of *techné*, the discourse that *carries* us *beyond* a purely physical and natural realm into another universe of discourse, turns back on itself to demonstrate just how "the stories we tell ourselves about ourselves," to use a phrase employed by Philip Glotzbach,[14] how our "representations" and "conceptions"—fictions, if you will—are already at work in the scene we wish to describe. Whatever the content, whatever is exemplified, whatever "object" is brought forth, discourse—and in this context, especially the discourse on *techné*—describes in advance the context and the conditions of its own description. The models or paradigms we create to provide explanatory power to our discourses are themselves facsimiles, rough copies for which there is no object or source except themselves. We work with and within a fabricated structure of iterability—reproduction within and across different systems of discourse.

One area of research in which questions of this nature play a predominant role is cognitive science and artificial intelligence (AI) research. In this realm, which one might first associate grossly with the generation of computer hardware and artificial languages, the question of technology's mediacy and ubiquity plays a dominating role. What more obvious exemplar is there of the artifice of *techné* than language—the "representation" of ourselves and our world to ourselves through the construction and use of signs, symbols, and metaphor. What better exemplar is there to indicate the mediacy and ubiquity of *techné*? Philip A. Glotzbach, in "Artificial Representationalism: Necessary (?) Constraints on Computer Models of Natural Language Understanding," and Lawrence Davis, in "From Artificial to Natural Intelligence: A Philosophical Critique," carry the discourse on the representation of human knowledge, the use and acquisition of language, and "natural intelligence" beyond their traditional parameters. They wish to examine critically the break-throughs and problems that AI research brings to the study of the relation between perception, language use, and representation.

Glotzbach traces a series of problems central to attempts in cognitive science and AI research to produce "propositional, computational" accounts of problem solving, natural language understanding, perception, and knowledge representation. His attention is directed toward the artificial constraints, a myopia of sorts, engendered by computational representation systems. Glotzbach argues that the tradition on which these systems are built, and which has dominated the philosophy of mind and psychology since at least Descartes, and perhaps even as far back in the history of philosophy as Plato, "consistently looked upon mental events and states as particulars existing entirely in the mind."[15] Such *representationalism* functions on the assumption that mental states correspond with specific states, events, and circumstances external to the mind; our ideas and concepts are thought to provide images of the external world. In an attempt to go beyond the limitations inherent to representationalism, and other problems evidenced in the writings of Fodor, Katz, Dennett, and others, Glotzbach introduces a way by which nonpropositional, nonrepresentational, and preconceptual elements of perception might be incorporated into AI research. Such a refashioning of artificial systems would have to take into account the "ecological approach to perception" as developed by James J. Gibson and a "phenomenology of lived-experience" articulated by Maurice Merleau-Ponty.

Lawrence Davis offers a critical philosophical survey of recent AI research to support his contention that the "evolution of our knowledge representation techniques is moving closer and closer to a 'true' theory of human thought."[16] Where Glotzbach introduces an "ecological" approach, Davis attempts to work out a "genetic" or "evolutionary" theory of knowledge representation and human thought as an alternative to the classical artificial systems that attempt to generate computational expert systems, or systems based on a logical calculus that views human thought as propositional, or again, representational in nature. According to Davis, "genetic algorithms" can be used to understand how humans "encode" and "decode" experience according to the specific environment in which they are situated. In this regard, Davis shares with Glotzbach the belief that AI research would benefit from the integration of an *evolutionary* account of human perceptual skills, skills to encode or decode experience that enhance our fitness in our environment.

The diversity of thematic focus presented in each essay traverses the heterogeneity of technology as a question. No presumption is made regarding the definition, the comprehension, or the interpretation of the term, the concept, or the equipment called "technology." Instead, every attempt is made to reflect the interlacing of meanings and directions by which our understanding of technology and ourselves comes to be what it is. These essays reflect not only the heterogeneity of technology's ubiquity, but they draw attention to, and draw on the need for, an interdisciplinary approach in the investigation of the question of technology. More specifically, the volume itself indicates the need to take into account, to give an account of, the different "levels" of understanding already at work in the engagement and study of technology. What these essays share, then, is the desire to grasp how, and to what extent, the emergence of human thought and activity in education, economics, politics, the arts and humanities, and the sciences is configured by the systematic deployments and manifestations of technology in our daily lives. Further, they share in the attempt to bring into focus the idea that our understanding of the world, in terms of our engagements, is mediated by technological concepts and practices. Moreover, these essays demonstrate that the knowledge we claim to possess about the world, the universe, our very existence (perhaps, especially, in our attempts to fix the furture) is given to us, *disclosed* as it were, through the very artifice and techniques—ideas, concepts, metaphors, and strategies of interpretation—used to fashion and articulate this knowledge.

Tracing a line of thought articulated by Plato as well as Hegel, Martin Heidegger notes that "when thinking attempts to pursue something that has claimed its attention it may happen that on the way it undergoes a change."[17] The relevance of Heidegger's remark can be seen in our attempts to articulate the identity of technology, to define technology. However it is to be known, understood, or determined, technology remains amorphous, in spite of, and because of, its appearance as visible artifacts: it withdraws into the structures and processes of thought and culture; and in this withdrawal technology reappears differently always. It is this heterogeneity of techniques, of technology, that challenges and questions our understanding of ourselves and technology. So rather than dwelling on the *content*—the *object* supposed in posing the question "What is technology?"—the essays that compose *From Artifact to Habitat* draw attention to the ways of being-with technology, or "the paths" thought pursues in an attempt to grasp how technology provides us with a means, a mode for translating the role it plays in the constitution of the medium or the habitat for human existence, the community. Thus, not only do these essays present a series of reflective and critical studies on the question of technology, but they suggest ways by which we can move toward a more commanding use and comprehension of the phenomenon and equipment we call "technology."

NOTES

1. Ludwig Wittgenstein, *Philosophical Investigations*, trans. G. E. M. Anscombe (New York: Macmillan, 1968), sec. 128.
2. Joseph Agassi, *Technology: Philosophical and Social Aspects* (Dordrecht: D. Reidel Publishing Company, 1985), p. 11. See also Herbert Marcuse, *One-Dimensional Man* (Boston: Beacon Press, 1964). Marcuse writes that "the prevailing forms of social control are technological" (p. 9).
3. See p. 99, below.
4. See Edmund Husserl, *Ideas: General Introduction to Pure Phenomenology*, trans. W. R. Boyce Gibson (New York: Collier-Macmillan, 1962), secs. 27–30; one should compare Husserl's discussion with David Hume's analysis of "vulgar systems" in *A Treatise of Human Nature*, 2d ed., ed. L. A. Selby-Bigge (Oxford: Oxford University Press, 1979), pp. 143–55, 181–262.
5. James L. Wiser, "Technological Consciousness and the Modern Understanding of the Good Life," see pp. 160–61 below.
6. See Jacques Ellul, *The Technological Society*, trans. John Wilkinson (New York: Alfred A. Knopf, 1964), esp. pp. 4–5 and pp. 80–83.
7. Langdon Winner, *Autonomous Technology: Technics-Out-of-Control as a Theme in Political Thought* (Cambridge: MIT Press, 1977); see esp. pp. 13–44, 279–305, and 306–35.
8. Langdon Winner, *The Whale and the Reactor: The Search for Limits in an Age of High Technology* (Chicago: University of Chicago Press, 1986), pp. 3–18.
9. See p. 132 below.
10. Ibid.
11. See pp. 166–72 below; and see Jean-Paul Sartre, *Critique of Dialectical Reason*, trans. Alan Sheridan-Smith, ed. Jonathan Reé (London: NLB, 1976), esp. pp. 253ff., 345–404, 445–504, 446–576, and 576–663.
12. Cf. Ellul, *The Technological Society*, pp. 1–7.
13. See Christopher T. Hill and James W. Utterback, *Technological Innovation for a Dynamic Economy* (New York: Pergamon Press, 1979) and Samuel Florman, *Blaming Technology: The Irrational Search for Scapegoats* (New York: St. Martin's Press, 1981).
14. See p. 175 below.
15. See p. 188 below.
16. See p. 196 below.
17. Martin Heidegger, "The Principle of Identity," *Identity and Difference*, trans. Joan Stambaugh (New York: Harper and Row, 1969), p. 23.

Contributors

FREDERIC L. BENDER is professor and chairperson of philosophy at the University of Colorado at Colorado Springs. His interest in technology arises from his work in social and political philosophy, phenomenology, and other related topics. Author of numerous articles on Marx, contemporary Marxism, Heidegger, and Merleau-Ponty, Professor Bender has edited *Karl Marx: The Essential Writings* (1986) and *The Communist Manifesto: A Norton Critical Edition* (1988).

EDMUND F. BYRNE teaches philosophy at Indiana University in Indianapolis where he holds the position of professor. Professor Byrne has published widely in the areas of the philosophy of technology and the philosophy of work. He has published two books, *Probability and Opinion* (1968) and (with Edward A. Maziarz) *Human Being and Being Human: Man's Philosophies of Man* (1969). He is co-editor (with Joseph C. Pitt) of *Technological Transformation*, a forthcoming publication in the Philosophy and Technology series. A book on *Work and Justice* is in the final stages of completion.

LAWRENCE DAVIS received his Ph.D. in philosophy from the University of Massachusetts at Amherst. Today he works at Bolt Beranek and Newman in Boston as a computer software specialist. Davis has published widely in the areas of logic, philosophy of language, and artificial intelligence. He is the editor of *Genetic Algorithms and Simulated Annealing* (1987).

PHILIP A. GLOTZBACH is associate professor and chairperson of the Department of Philosophy at Denison University. He has published in the areas of the philosophy of psychology and the philosophy of language. His primary research interests are perception theory and philosophical issues in artificial intelligence. Glotzbach is currently completing a book on J. J. Gibson's psychological theory of perception.

DAVID LOVEKIN teaches philosophy at Hastings College where he is an associate professor. He has published extensively on the thought of Jacques Ellul and Giambattista Vico, including "Jacques Ellul and the Logic of Technology," "Giambattista Vico and Jacques Ellul: The Intelligible Consciousness and the Technological Phenomenon," and "Technology as the Sacred Order." Lovekin is completing a book entitled *Technique, Discourse, and Consciousness: An Introduction to the Thought of Jacques Ellul.*

CARL MITCHAM is director of the Philosophy & Technology Studies Center at Polytechnic University in New York. Known for his work on the bibliographies of current research in science and technology studies and the philosophy of technology, Mitcham is the author of numerous articles, including "Types of Technology" and "Philosophy and the History of Technology." Moreover, he is the editor and author of several major books in the field, including *Bibliography of the Philosophy of Technology* (with Robert Mackey, 1973), *Philosophy and Technology* (edited with Robert Mackey, 1972), *Technology and Theology: Essays in Christian Analysis and Exegesis* (edited with Jim Grote, 1984), and most recently, *Philosophy and Technology II: Information Technology and Computers in Theory and Practice* (with Alois Huning, 1986).

GAYLE L. ORMISTON is associate professor in the Department of Philosophy and the Institute for Applied Linguistics at Kent State University. In addition to published articles on issues related to representation, language, and philosophical critique in the work of Nietzsche, Heidegger, and Derrida, he has edited (with Alan D. Schrift) *The Hermeneutic Tradition: From Ast to Ricoeur* (1989) and *Transforming the Hermeneutical Context: From Nietzsche to Nancy* (1989). *Narrative Experiments: The Discursive Authority of Science and Technology*, co-authored with Raphael Sassower, and *Prescriptions: The Dissemination of Medical Authority*, coedited with Raphael Sassower, are also forthcoming.

JOSEPH C. PITT is professor of philosophy at the Virginia Polytechnic Institute and State College where he is director of the Humanities, Science, and Technology Program. Author of numerous articles in the philosophy of science and the philosophy of technology, Pitt has authored and edited nine books, including *Theories of Explanation* (1988), *Rational Changes in Science* (1988), and *Change and Progress in Modern Science* (1985). His introductory text to *The Philosophy of Technology* is forthcoming.

JAMES L. WISER is professor of political science at Loyola University of Chicago. His research and teaching interests are in the area of political philosophy. Among his publications are *Political Philosophy: A History of the Search for Order* (1983) and *Political Theory: A Thematic Inquiry* (1986).

From Artifact to Habitat

Three Ways of Being-With Technology

CARL MITCHAM

In any serious discussion of issues associated with technology and humanity there readily arises a general question about the primary member in this relationship. On the one hand, it is difficult to deny that we exercise some choice over the kinds of technics with which we live—that is, that we control technology. On the other, it is equally difficult to deny that technics exert profound influences on the ways we live—that is, structure our existence. We build our buildings, Winston Churchill once remarked (apropos a proposal for a new Parliament building), then our buildings build us. But which comes first, logically if not temporally, the builder or the buildings? Which is primary, humanity or technology?

This is, of course, chicken-and-egg question, one not subject to any straightforward or unqualified answer. But it is not therefore insignificant, nor is it enough to propose as some kind of synthesis that there is simply a mutual relationship between the two, that humanity and technology are always found together. Mutual relationship is not some one thing; mutual relationships take many different forms. There are, for instance, mutualities of parent and child, of husband and wife, of citizens, and so forth. Humanity and technology can be found together in more than one way. Rather than argue the primacy of one or the other factor or the cliché of mutuality in the humanity-technology relationship, I propose to outline three forms the relationship itself can take, three ways of being-with technology.

To speak of three ways of being-with technology is necessarily to borrow and adapt a category from Martin Heidegger's *Being and Time* (1927) in a manner that deserves some acknowledgment. In his seminal work Heidegger proposes to develop a new understanding of being human by taking the primordial human condition,

31

being-in-the-world, and subjecting this given to what he calls an existential analysis. The analysis proceeds by way of elucidating three equiprimordial aspects of this condition of being human: *the world* within which the human finds itself, the *being-in* relationship, and the *being* who is in the relationship—all as a means of approaching what, for Heidegger, is the fundamental question, the meaning of Being.

The fundamental question need not, on this occasion, concern us. What does concern us is the central place of technics in Heidegger's analysis and the disclosure of being-with as one of its central features. For Heidegger the worldhood of the world, as he calls it, comes into view through technical engagements, which reveal a network of equipment and artifacts ready-to-hand for manipulation, and other human beings likewise so engaged. These others are neither just technically ready-to-hand (like tools) nor even scientifically present-at-hand (like natural objects); on the contrary, they are *like* the very human being who notices them in that "*they are there too, and there with it.*"[1]

The being-with relationship thus disclosed through technical engagements is, for Heidegger, primarily social in character; it refers to the social character of the world that comes to light through technical practice. Such a world is not composed solely of tools and artifacts, but of tools used with others, and artifacts belonging to others. Technical engagements are not just technical but have an immediately and intimately social dimension. Indeed, this is all so immediate that it requires some labored stepping back even to recognize and state—the processes of distancing and articulation that are in part precisely what philosophy is all about.

The present attempt to step back and examine various ways of being-with technology rather than being-with others (through technology) takes off from but does not proceed in the same manner as Heidegger's social analysis of the They and the problem of authenticity in the technological world. For Heidegger, being-with refers to an immediate personal presence in technics. Social being-with can manifest itself, however, not just on the level of immediate or existential presence but also in ideas. Indeed, the social world is as much, if not more, a world of ideas as of persons. Persons hold ideas and interact with others and with things on the basis of them. These ideas can even enclose the realm of technics—that is, become a language or *logos* of technics, a "technology."

The idea of being-with technology presupposes this "logical" encompassing of technics by a society and its philosophical or protophilosophical articulation. For many people, however, the ideas that guide their lives may not be held with conscious awareness or full articulation. They often take the form of myth. Philosophical argument and discussion then introduces into such a world of ideas a kind of break or rupture with the immediately given. This break or rupture need not require the rejection or abandoning of that given, but it will entail the bringing of that given in to fuller consciousness or awareness—from which it must be accepted (or rejected) in a new way or on new grounds.

Against this background, then, I propose to develop historico-philosophical descriptions, necessarily somewhat truncated, of three alternative ways of being-with technology. The first is what may be called ancient skepticism; the second, Renaissance and Enlightenment optimism; and the third, romantic ambiguity or uneasiness. Even in the somewhat simplified form of ideal types in which they will be presented here, consideration of the issues that divide these three ways of being-with technology may perhaps illuminate the difficulties we face in trying to live with modern technology and its manifest problems.

1

The original articulation of a relationship between humanity and technics, an articulation that is in its earliest forms coeval with the appearance of recorded history, can be stated boldly as "technology is bad but necessary" or, perhaps more carefully, as "technology (that is, the study of technics) is necessary but dangerous." The idea is hinted at by a plethora of archaic myths, such as the story of the Tower of Babel or the myths of Prometheus, Hephaestos, or Daedalus and Icarus. Certainly the transition from hunting and gathering to the domestication of animals and plants introduced a profound and profoundly disturbing transition into culture. Technics, according to these myths, although to some extent required by humanity and thus on occasion a cause for legitimate celebration,[2] easily turn against the human by severing it from some larger reality—a severing that can be manifest in a failure of faith or shift of the will, a refusal to rely on or trust God or the gods, whether manifested in nature or in providence.[3]

Ethical arguments in support of this distrust or uneasiness about technical activities can be detected in the earliest strata of Western philosophy. According to the Greak historian Xenophon, for instance, his teacher Socrates (469–399 B.C.) considered farming, the least technical of the arts, to be the most philosophical of occupa-

tions. Although the earth "provides the good things most abundantly, farming does not yield them up to softness but. . . produces
a kind of manliness. . . . Moreover, the earth, being a goddess,
teaches justice to those who are able to learn" (*Oeconomicus* 5.4
and 12). This idea of agriculture as the most virtuous of the arts, one
in which human technical action tends to be kept within proper
limits, is repeated by representatives of the philosophical tradition
as diverse as Plato,[4] Aristotle,[5] St. Thomas Aquinas,[6] and Thomas
Jefferson.[7]

Elsewhere Xenophon notes Socrates' distinction between questions about *whether* to perform an action and *how* to perform it,
along with another distinction between scientific or technological questions concerning the laws of nature and ethical or political
questions about what is right and wrong, good and bad, pious and
impious, just and unjust. In elaborating on the first distinction,
Socrates stresses that human beings must determine for themselves
how to perform their actions—that they can take lessons in "construction (*tektonikos*), forging metal, agriculture, ruling human
beings, and. . . calculation, economics, and military strategy," and
therefore should not depend on the gods for help in "counting,
measuring, or weighing"; the ultimate consequences of their technical actions are nonetheless hidden. His initial example is even
taken from agriculture: the man who knows how to plant a field
does not know whether he will reap the harvest. Thus whether we
should employ our technical powers is a subject about which we
must rely on guidance from the gods.[8]

At the same time, with regard to the second distinction, Socrates
argues that because of the supreme importance of ethical and political issues, human beings should not allow themselves to become
preoccupied with scientific and technological pursuits. In the intellectual autobiography attributed to him in the *Phaedo*, for instance,
Socrates relates how he turned away from natural science because of
the cosmological and moral confusion it tended to engender.[9] In the
Memorabilia it is similarly said of Socrates that

> He did not like others to discuss the nature of all things, nor did he
> speculate on the "cosmos" of the sophists or the necessities of the
> heavens, but he declared that those who worried about such matters
> were foolish. And first he would ask whether such persons became in-
> volved with these problems because they believed that their knowledge
> of human things was complete or whether they thought they were obli-
> gated to neglect human things to speculate on divine things. (*Memorabil-
> ia* 1.1.11–12)

Persons who turn away from human things to things having to do with the heavens appear to think "that when they know the laws by which everything comes into being, they will, when they choose, create winds, water, seasons, and anything else like these that they may need" (*Memorabilia* 1.1.15).[10] As "the first to call down philosophy from the heavens and place it in the city and . . . compel it to inquire about life and morality and things good and bad,"[11] Socrates' own conversation, however, is described as always about human things: What is pious? What is impious? What is good? What is shameful? What is just? What is unjust? What is moderation? For, as Xenophon says on another occasion, Socrates "was not eager to make his companions orators and businessmen and inventors, but thought that they should first possess moderation [*sophrosune*]. For he believed that without moderation those abilities only enabled a person to become more unjust and to work more evil[11] (*Memorabilia* 4.3). The initial distinction grants technical or "how-to" questions a realistic prominence in human affairs but recognizes their ambiguity and uncertainty; the subsequent distinction subordinates any systematic pursuit of technical knowledge to ethical and political concerns.

Such uneasiness before the immoderate possibilities inherent in technological powers is further elaborated by Plato. Near the beginning of the *Republic*, after Socrates outlines a primitive state and Glaucon objects that this is no more than a "city of pigs," Socrates replies

> The true state is in my opinion the one we have described—a healthy state, as it were. But if you want, we can examine a feverish state as well. . . . For there are some, it seems, who will not be satisfied with these things or this way of life; but beds, tables, and other furnishings will have to be added, and of course seasonings, perfume, incense, girls, and sweets—all kinds of each. And the requirements we mentioned before can no longer be limited to the necessities of houses, clothes, and shoes; but [various *technai*] must also be set in motion. . . . The healthy state will no longer be large enough either, but it must be swollen in size by a multitude of activities that go beyond the meeting of necessities. . . . (372d–373b)

As this passage indicates, and as can be confirmed by earlier references to Homer and other poets, certain elements of classical Greek culture had a distrust of the wealth and affluence that the *technai* or arts could produce if not kept within strict limitations. For according to the ancients such wealth accustoms men to easy things.

But *chalepa ta kala*, difficult is the beautiful or the perfect; the
perfection of anything, including human nature, is the opposite of
what is soft or easy. Under conditions of affluence human beings
tend to become accustomed to ease, and thus tend to choose the less
over the more perfect, the lower over the higher, both for themselves and for others. With no art is this more prevalent than with
medicine. Once drugs are available as palliatives, for instance, most
individuals will choose them for the alleviation of pain over the
more strenuous paths of physical hygiene or psychological enlightenment. The current (*techné*) of medicine, Socrates maintains to
Glaucon later in the third book of the *Republic*, is an education in
disease that "draws out death" (406b); instead of promoting health
it allows the unhealthy to have "a long and wretched life" and "to
produce offspring like themselves" (407d). That Socrates' description applies even more strongly to modern medical technology than
it did to that in classical Athens scarcely needs to be mentioned.

Another aspect of this tension between politics and technology
is indicated by Plato's observations on the dangers of technical
change. In the words of Adeimantus, with whom Socrates in this
instance evidently agrees, once change has established itself as normal in the arts, "it overflows its bounds into human character and
activity and from there issues forth to attach commerical affairs, and
then proceeds against the laws and political orders" (424d–e). It is
desirable that obedience to the law should rest primarily on habit
rather than force. Technological change, which undermines the authority of custom and habit, thus tends to introduce violence into
the state. Surely this is a possibility that the experience of the twentieth century, one of the most violent in history, should encourage
us to take seriously.

This wariness of technological activity on moral and political
grounds can be supplemented by an epistemological critique of the
limitations of technological knowledge and a metaphysical analysis
of the inferior status of technical objects. During a discussion of the
education of the philosopher-king in the seventh book of the *Republic*, Socrates considers what kind of teaching most effectively brings a
student "into the light" of the highest or most important things. One
conclusion is that it is not those *technai* that "are oriented toward
human opinions and desires or concerned with creation and fabrication and attending to things that grow and are put together" (533b).
Because it cannot bring about a conversion or emancipation of the
mind from the cares and concerns of the world, technology should
not be a primary focus of human life. The orientation of technics,
because it is concerned to remedy the defects in nature, is always

toward the lower or the weaker (342c–d). A physician sees more sick people than he does healthy ones. *Eros* or love, by contrast, is oriented toward the higher or the stronger; it seeks out the good and strives for transcendence. "And the person who is versed in such matters is said to have spiritual wisdom, as opposed to the wisdom of one with *technai* or low-grade handicraft skills," Diotima tells Socrates in the *Symposium* (203a).

Aristotle agrees, but for quite different, more properly metaphysical reasons. According to Aristotle and his followers, reality or being resides in particulars. It is not some abstract species Homo Sapiens (with capital H and capital S) that *is* in the primary sense, but Socrates and Xanthippe. However, the reality of all natural entities is dependent on an intimate union of form and matter, and the *telos* or end determined thereby. The problem with artifacts is that they fail to achieve this kind of unity at a very deep level, and are thus able to have a variety of uses or extrinsic ends imposed upon them. "If a bed were to sprout," says Aristotle, "not a bed would come up but a tree" (*Physics* 193b.10). Insofar as it truly imitates nature, art engenders an inimitable individuality in its products, precisely because its attempt to effect as close a union of form and matter as possible requires a respect for or deference to the materials with which it works. In a systematized art or technology, matter necessarily tends to be overlooked or relegated to the status of an undifferentiated substrate to be manipulated at will.[12] Indeed, in relation to this Aristotle suggests a distinction between the arts of cultivation (e.g., medicine, education, and agriculture, which help nature to produce more abundantly things that it can produce of itself) from those of construction or domination (arts that bring into existence things that nature would not).[13]

The metaphysical issue here can be illustrated by observing the contrast between a handcrafted ceramic plate and Tupperware dishes. The clay plate has a solid weight, rich texture, and explicit reference to its surroundings not unlike that of a natural stone, whereas Tupperware exhibits a lightness of body and undistinguished surface that only abstractly engages the environment of its creation and use. According to a Mobil Oil Company advertisement from the early 1980s, synthetic products are actually "better than the real thing," so the word "synthetic," which implies a "pallid imitation," ought to be discarded. But whether this is true or not depends heavily on a prior understanding of what is real in the first place. For Aristotle there is a kind of reality that can only be found in particulars and is thus beyond the grasp of mass-production, function-oriented, polymer technology.

For Plato and the Platonic tradition, too, artifice is less real than nature. Indeed, in the tenth book of the *Republic* there is a discussion of the making of beds (to which Aristotle's remarks from the *Physics* may allude) by god or nature, by the carpenter or *tekton*, and by painter or artist. Socrates' argument is that the natural bed, the one made by the god, is the primary reality; the many beds made in imitation by artisans are a secondary reality; and the pictures of beds painted by artists are a tertiary reality. *Techné* is thus creative in a second or "third generation" sense (597e)—and thus readily subject to moral and metaphysical guidance.

In moral terms artifice is to be guided or judged in terms of its goodness or usefulness. In metaphysical terms the criterion of judgment is proper proportion or beauty. One possible disagreement between Platonists and Aristotelians with regard to one or another aspect of making is whether the good or the beautiful, ethics or aesthetics, is the proper criterion for its guidance. Such disagreement should nevertheless not be allowed to obscure a more fundamental agreement, the recognition of the need to subject *poesis* and *technai* to certain well-defined limitations. Insofar as technical objects or activities fail to be subject to the inner guidance of nature (*phusis*), nature must be brought to bear upon them consciously, from the outside (as it were) by human beings. Again, the tendency of contemporary technical creations to bring about environmental problems or ecological disorders to some extent confirms the premodern point of view.

The ancient critique of technology thus rests on a tightly woven, fourfold argument: (1) the will to technology or the technological intention often involves a turning away from faith or trust in nature or providence; (2) technical affluence and the concomitant processes of change tend to undermine individual striving for excellence and societal stability; (3) technological knowledge likewise draws the human being into intercourse with the world and obscures transcendence; (4) technical objects are less real than objects of nature. Only some necessity of survival, not some ideal of the good, can justify the setting aside of such arguments. The life of the great Hellenistic scientist Archimedes provides us (as it did antiquity) with a kind of icon or lived-cut image of these arguments. Although, according to Plutarch, Archimedes was capable of inventing all sorts of devices, he was too high minded to do so except when pressed by military necessity—yet even then he refused to leave behind any treatise on the subject, because of a salutary fear that his weapons would be too easily misused by humankind.[14]

Allied with the Judeo-Christian-Islamic criticism of the vanity of human knowledge and of worldly wealth and power,[15] this premodern distrust of technology dominated Western culture until the end of the Middle Ages, and elements of it can be found vigorously re-peated in numerous figures since—from Samuel Johnson's neo-classicist criticism of Milton's promotion of edcuation in natural science[16] to Norbert Wiener who in 1947, like Archimedes twenty-three hundred years before, vowed not to publish anything more that could do damage in the hands of militarists.[17] In one less well-known allusion to another aspect of the classical moral argument, John Wesley (1703–91), in both private journals and public ser-mons, ruefully acknowledges the paradox that Christian conversion gives birth to a kind of self-discipline that easily engenders the accu-mulation of wealth—which wealth then readily undermines true Christian virtue. "Indeed, according to the natural tendency of riches, we cannot expect it to be otherwise," writes Wesley.[18]

With regard to other aspects of the premodern critique, Lewis Mumford, for instance, has criticized the will to power manifested in modern technology, and Heidegger, following the lead of the poet Ranier Maria Rilke, has invoked the metaphysical argument by pointing out the disappearance of the thinghood of things, the loss of a sense of the earth in mass-produced consumer objects. From Heidegger's point of view, nuclear annihilation of all things would be "the mere final emission of what has long since taken place, has already happened."[19]

From the viewpoint of the ancients, then, being-with technology is an uneasy being-along-side-of and working-to-keep-at-arms-length. Phrased in terms of the contemporary discipline of technolo-gy assessment, this premodern attitude looks upon technics as dangerous or guilty until proven innocent or necessary—and in any case, the burden of proof lies with those who would favor technolo-gy, not those who would restrain it.

2

A radically different way of being-with technology—one that shifts the burden of proof from those who favor to those who oppose the introduction of inventions—argues the inherent goodness of technology and the consequent accidental character of all misuse. Aspects of this idea or attitude are not without premodern adum-bration. But in comprehensive and persuasive form arguments to this effect are first fully articulated in the writings of Francis Bacon (1561–1626) at the time of the Renaissance, and subsequently become characteristic of the Enlightenment philosophy of the eighteenth century.

Like Xenophon's Socrates, Bacon grants that the initiation of human actions should be guided by divine counsel. But unlike Socrates, Bacon maintains that God has given humanity a clear mandate to pursue technology as a means for the compassionate alleviation of the suffering of the human condition, of being-in-the-world. Technical know-how is cut loose from all doubt about the consequences of technical action. In the choice between ways of life devoted to scientific-technological or ethical-political questions, Bacon further argues that Christian revelation directs men toward the former over the latter.

> For it was not that pure and uncorrupted natural knowledge whereby Adam gave names to the creatures according to their propriety, which gave occasion to the fall. It was the ambitious and proud desire of moral knowledge to judge of good and evil, to the end that man may revolt from God and give laws to himself, which was the form and manner of the temptation.[20]

Contrary to what is implied by the myth of Prometheus or the legend of Faust, it was not scientific and technological knowledge that led to the Fall, but vain philosophical speculation concerning moral questions. Formed in the image and likeness of God, human beings are called upon to be creators; to abjure that vocation and to pursue instead an unproductive discourse on ethical quandaries brings about the just punishment of a poverty-stricken existence. "He that will not apply new remedies must expect new evils."[21] Yet "the kingdom of man, founded on the sciences," says Bacon, is "not much other than . . . the kingdom of heaven."[22]

The argument between Socrates and Bacon is not, it is important to note, simply one between partisans either against or in favor of technology. Socrates allows technics a legitimate but strictly utilitarian function, then points out the difficulty of obtaining a knowledge of consequences upon which to base any certainty of trust or commitment. Technical action is circumscribed by uncertainty or risk. Bacon, however, although he makes some appeals to a consequentialist justification, ultimately grounds his commitment in something approaching deontological principles. The proof is that he never even considers evaluating individual technical projects on their merit, but simply argues for an all-out affirmation of technology in general. It is right to pursue technological action, never mind the consequences. Intuitions of uncertainty are jettisoned in the name of revelation.

The uniqueness of the Baconian (or Renaissance) interpretation of the theological tradition is also to be noted. For millennia the doctrines of God as creator of "the heavens and the earth" (Gen. 1.1) and human beings as made "in the image of God" (Gen. 1.27) exercised profound influence over Jewish and later Christian anthropology, without ever being explicitly interpreted as a warrant for or a call to technical activity. Traditional or premodern interpretations focus on the soul, the intellect, or the capacity for love as the key to the *imago Dei*.[23] The earliest attribution to this doctrine of technological implications occurs in the early Renaissance.[24] The contemporary theological notion of the human as using technology to prolong creation or cocreate with God depends precisely on the reinterpretation of Genesis adumbrated by Bacon.

The Enlightenment version of Bacon's religious argument is to replace the theological obligation with a natural one. In the first place, human beings simply could not survive without technics. As D'Alembert puts it in the "Preliminary Discourse" to the *Encyclopedia* (1751), there is a prejudice against the mechanical arts that is a result of their accidental association with the lower classes.

> The advantage that the liberal arts have over the mechanical arts, because of their demands upon the intellect and because of the difficulty of excelling in them, is sufficiently counterbalanced by the quite superior usefulness which the latter for the most part have for us. It is their very usefulness which reduced them perforce to purely mechanical operations in order to make them accessible to a larger number of men. But while justly respecting great geniuses for their enlightenment, society ought not to degrade the hands by which it is served.[25]

In the even more direct words of Immanuel Kant, "Nature has willed that man should by himself, produce everything that goes beyond the mechanical ordering of his animal existence, and that he should partake of no other happiness or perfection than that which he himself, independently of instinct, has created by his own reason."[26] Nature and reason, if not God, command humanity to pursue technology; the human being is redefined not as *homo sapiens* but as *homo faber*. Technology is the essential human activity. In more ways than Kant explicitly proclaims, "Enlightenment is man's release from his self-incurred tutelage."[27]

Following a redirecting (Bacon) or reinterpreting (D'Alembert and Kant) of the will, Bacon and his followers explicitly reject the ethical-political argument against technological activities in the name of moderation. With no apparent irony, Bacon maintains that the inventions of printing, gunpowder, and the compass have done

more to benefit humanity than all the philosophical debates and
political reforms throughout history. It may, he admits, be perni-
cious for an individual or a nation to pursue power. Individuals or
small groups may well abuse such power. "But if a man endeavor to
establish and extend the power and dominion of the human race
itself over the universe," writes Bacon, "his ambition (if ambition it
can be called) is without doubt both a more wholesome and a more
noble thing than the other two." And, of course, "the empire of
man over things depends wholly on the arts and sciences." (*Novum
Organum* 1.129).

Bacon does not expound at length on the wholesomeness of technics. All he does is reject the traditional ideas of their corrupting
influence on morals by arguing for a distinction between change in
politics and in the arts.

> In matters of state a change even for the better is distrusted [Bacon
> observes], because it unsettles what is established; these things resting
> on authority, consent, fame and opinion, not on demonstration. But arts
> and sciences should be like mines, where the noise of new works and
> further advances is heard on every side. (*Novum Organum* 1.90)

Unlike Aristotle and Aquinas, both of whom noticed the same
distinction but found it grounds for caution in technology,[28] Bacon
thinks the observation itself is enough to set technology on its own
path of development.

Bacon's Enlightenment followers, however, go considerably
further, and argue for the positive or beneficial influence of the arts
on morals. The *Encyclopedia*, for instance, having identified "luxury" as simply "the use human beings make of wealth and industry
to assure themselves of a pleasant existence" with its origin in "that
dissatisfaction with our condition . . . which is and must be present in
all men," undertakes to reply directly to the ancient "diatribes by
the moralists who have censured it with more gloominess than
light."[29] Critics of material welfare have maintained that it undermined morals, and apologists have responded that this is the case
only when it is carried to excess. Both are wrong. Wealth is, as we
would say today, neutral. A survey of history reveals that luxury
"did not determine morals, but . . . it took its character rather from
them." [30] Indeed, it is quite possible to have a moral luxury, one
that promotes virtuous development.

But if a first line of defense is to argue for moderation, and a
second for neutrality, a third is to maintain a positive influence.
David Hume (1711–76), for instance, in his essay "Of Commerce,"
argues that a state should encourage its citizens to be manufacturers

more than farmers or soldiers. By pursuit of "the arts of *luxury*, they add to the happiness of the state."[31] Then, in "Of Refinement in the Arts," he explains that the ages of luxury are both "the happiest and the most virtuous" because of their propensity to encourage industry, knowledge, and humanity."[32] "In times when industry and the arts flourish," writes Hume, "men are kept in perpetual occupation, and enjoy, as their reward, the occupation itself, as well as those pleasures which are the fruit of their labour."[33]

Furthermore, the spirit of activity in the arts will galvanize that in the sciences and vice versa; knowledge and industry increase together. In Hume's own inimitable words: "We cannot reasonably expect that a piece of woolen cloth will be wrought to perfection in a nation which is ignorant of astronomy."[34] And the more the arts and sciences advance, "the more sociable men become." Technical engagements promote civil peace because they siphon off energy that might otherwise go into sectarian competition. Technological commerce and scientific aspirations tend to break down national and class barriers, thus ushering in tolerance and sociability. In the words of Hume's contemporary, Montesquieu: "Commerce is a cure for the most destructive prejudices; for it is a general rule, that wherever we find tender manners, there commerce flourishes; and that wherever there is commerce, there we meet with tender manners."[35]

The ethical significance of technological activity is not limited, however, to its socializing influence. Technology is an intellectual as well as a moral virtue, because it is a means to the acquisition of true knowledge. That technological activity contributes to scientific advance rests on a theory of knowledge that again is first clearly articulated by Bacon, who begins his *Novum Organum* or "new instrument" with the argument that true knowledge is acquired only by a close intercourse with things themselves. "Neither the naked hand nor the understanding left to itself can effect much. It is by instruments and helps that the work is done, which are as much wanted for the understanding as for the hand" (*Novum Organum* 1.2). Knowledge is to be acquired by active experimentation, and ultimately evaluated on the basis of its ability to engender works. The means to true knowledge is what Bacon candidly refers to as the "torturing of nature"; left free and at large, nature, like the human being, is loath to reveal her secrets.[36] The result of this new way will be the union of knowledge and power (*Novum Organum* 2.3). Bacon is, quite simply, an epistemological pragmatist. What is true is what works. "Our only hope," he says, "therefore lies in a true induction" (*Novum Organum* 1.14).

The very basis of the great French *Encyclopedia or Dictionary of*

Sciences, Arts, and Crafts is precisely this epistemological vision, a
unity between theory and practice. Bacon is explicitly identified as
its inspiration, and is praised for having conceived philosophy "as
being only that part of our knowledge which should contribute to
making us better or happier, thus . . . confining it within the limits of
useful things [and inviting] scholars to study and perfect the arts,
which he regards as the most exalted and most essential part of human science."[37] Indeed, in explicating the priorities of the *Encyclopedia*, the "Preliminary Discourse" goes on to say that "too much
has been written on the sciences; not enough has been written well
on the mechanical arts."[38] The article on "Art" in the *Encyclopedia*
further criticizes the prejudice against the mechanical arts, not only
because it has "tended to fill cities with . . . idle speculation,"[39] but
even more because of its failure to produce genuine knowledge. "It
is difficult if not impossible . . . to have a thorough knowledge of the
speculative aspects of an art without being versed in its practice,"
although it is equally difficult "to go far in the practice of an art
without speculation."[40] It is this new unity of theory and practice—a
unity based in practice more than in theory[41]—that is at the basis of,
for instance, Bernard de Fontenelle's eulogies of the practice of experimental science as an intellectual virtue as well as a moral one,
and the Enlightenment reconception of Socrates as having called
philosophy down from the heavens to experiment with the world.[42]

Bacon's true induction likewise rests on a metaphysical rejection
of natural teleology. The pursuit of a knowledge of final causes
"rather corrupts than advances the sciences," declares Bacon, "except such as have to do with human action" (*Novum Organum* 2.2).
Belief in final causes or purposes inherent in nature is a result of
superstition or false religion. It must be rejected in order to make
possible "a very diligent dissection and anatomy of the world"
(*Novum Organum* 1.124). Nature and artifice are not ontologically
distinct. "All Nature is but Art, unknown to thee," claims Alexander Pope.[43] "Nature does not exist," declares Voltaire, "art is everything." The Aristotelian distinction between arts of cultivation
and of construction is jettisoned in favor of universal construction.

With regard to Pope, although it is not uncommon to find com-
parisons of the God/nature and artist/artwork relationships in Greek
and Christian, ancient and modern authors, there are subtle differences. For Plato (*Sophist* 265b and *Timaeus* 27c) and St. Augustine
(*De civitate Dei* 11.21), for example, there is first a fundamental
distinction to be made between divine and human *poiesis*, which is
itself to be distinguished from *techné* and, second, the fact that even
though made by a god the world is not to be looked upon as an

artifact or something that functions in an artificial manner. Thomas Hobbes, Bacon's secretary, however, proposes to view nature not just as produced by a divine art but as itself "the art whereby God hath made and governs the world."[44] Indeed, so much is this the case that for Hobbes human art itself can be said to produce natural objects—or, to say the same thing in different words, the whole distinction between nature and artifice disappears.

This last point also links up with the first: metaphysics supports volition. If nature and artifice are not ontologically distinct, then the traditional distinction between technics of cultivation and technics of domination disappears. There are no technics that help nature to realize its own internal reality, and human beings are free to pursue power. If nature is just another form of mechanical artifice, it is likewise reasonable to think of the human being as a machine. "Man is a machine and . . . in the whole universe there is but a single substance variously modified," concludes LaMettrie.[45] "For what is the heart," wrote Hobbes a century earlier, "but a spring; and the nerves, but so many strings; and the joints, but so many wheels."[46] But the activities appropriate to machines are technological ones; *homo faber* is yet another form of *l'homme machine*, and vice versa.

Like that of the anicents, then, the distinctly modern way of being-with technology may be articulated in terms of four interrelated arguments: (1) the will to technology is ordained for humanity by God or by nature; (2) technological activity is morally beneficial because, while stimulating human action, it ministers to physical needs and increases sociability; (3) knowledge acquired by a technical closure with the world is truer than abstract theory; and (4) nature is no more real than—indeed it operates by the same principles as—artifice. It is scarcely necessary to illustrate how aspects of this ideology remain part of intellectual discourse in Marxism, in pragmatism, and in popular attitudes regarding technological progress, technology assessment and public policy, education, and medicine.

3

The premodern argument that technology is bad but necessary characterizes a way of being-with technology that effectively limited rapid technical expansion in the West for approximately two thousand years. The Renaissance and Enlightenment argument in support of the theory that technology is inherently good discloses a way of being-with technology that has been the foundation for a

Promethean unleashing of technical power unprecedented in history. The proximate causes of this radical transformation were, of course, legion: geographic, economic, political, military, scientific. But what brought all such factors together in England in the mid-eighteenth century to engender a new way of life, what enabled them to coalesce into a veritable new way of being-in-the-world, was a certain kind of optimism regarding the expansion of material development that is not to be found so fully articulated at any point in premodern culture.[47]

In contrast with premodern skepticism about technology, however, the typically modern optimism has not retained its primacy in theory even though it has continued to dominate in practice. The reasons for this are complex. But faced with the real-life consequences of the Industrial Revolution, from societal and cultural disruptions to environmental pollution, post-Enlightenment theory has become more critical of technology. Romanticism, as the name for the typically modern response to the Enlightenment, thus implicitly contains a new way of being-with technology, one that can be identified with neither ancient skepticism nor modern optimism.

Romanticism is, of course, a multi-dimensioned phenomenon. In one sense, it can refer to a permanent tendency in human nature that manifests itself differently at different times. In another, it refers to a particular manifestation in nineteenth-century literature and thought. Virtually all attempts to analyze this particular historical manifestation interpret romanticism as a reaction to and criticism of modern science. Against Newtonian mechanics, the romantics propose an organic cosmology; in opposition to scientific rationality, romantics assert the legitimacy and importance of imagination and feeling. What is seldom appreciated is the extent to which romanticism can also be interpreted as a questioning—in fact, the first self-conscious questioning—of modern technology.[48] So interpreted, however, romanticism reflects an uneasiness about technology that is nevertheless fundamentally ambiguous; although as a whole the romantic critique may be distinct from ancient skepticism and modern optimism, in its parts it nevertheless exhibits differential affinities with both.

To begin with, consider the volitional aspect of technology. On the ancient view, technology was seen as a turning away from God or the gods. On the modern, it is ordained by God or, with the Enlightenment rejection of God, by Nature. With the romantics the will to technology either remains grounded in nature or is cut free from all extrahuman determination. In the former instance, however, nature is reconceived as not just mechanistic movement but as an organic striving toward creative development and expression. From the perspective of "mechanical philosophy," human technology is a prolongation of mechanical order; from that of *Naturphilosophie* it becomes a participation in the self-expression of life. When liberated from even such organic creativity, technology is grounded solely in the human will to power, but with recognition of its often negative consequences; the human condition takes on the visage of Gothic pathos.[49] The most that seems able to be argued is that the technological intention, that is the will to power, should not be pursued to the exclusion of other volitional options—or that it should be guided by aesthetic ideals.

William Wordsworth (1770–1850), for instance, the most philosophical of the English romantic poets, in the next-to-last book of his long narrative poem, *The Excursion* (1814), describes how he has "lived to mark / A new and unforseen creation rise" (bk. 8, lines 89–90).

> Casting reserve away, exult to see
> An intellectual mastery exercised
> O'er the blind elements; a purpose given,
> A perseverance fed; almost a soul
> Imparted—to brute matter. I rejoice,
> Measuring the force of those gigantic powers
> That, by the thinking mind, have been compelled
> To serve the will of feeble-bodied Man.
>
> (bk. 8, lines 200–207)

Here the rejoicing in and affirmation of technological conquest and control is clearly in harmony with Enlightenment sentiments.

> Yet in the midst of this exultation
> I grieve, when on the dark side
> Of this great change I look; and there behold
> Such outrage done to nature.
>
> (151–53)

And afterward he writes,

> How insecure, how baseless in itself,
> Is the Philosophy whose sway depends
> On mere material instruments;—how weak
> Those arts, and high inventions, if unpropped
> By virtue.
>
> (223–27)

Here Enlightenment optimism is clearly replaced by something approaching premodern skepticism.

Wordsworth clarifies his position in the last book of the poem. True, he has complained, in regard to the factory labor of children, that a child is

> . . . subject to the arts
> Of modern ingenuity, and made
> The senseless member of a vast machine.
>
> (bk. 9, lines 157–59)

Still, he is not insensitive to the fact that the rural life is also often an "unhappy lot" enslaved to "ignorance" "want" and "miserable hunger" (lines 163–65). Nevertheless, he says, his thoughts cannot help but be

> . . . turned to evils that are new and chosen,
> A bondage lurking under shape of good,—
> Arts, in themselves beneficent and kind,
> But all too fondly followed and too far.
>
> (187–90)

In such lines Wordsworth no longer maintains with any equanimity the Enlightenment principle that the arts are "in themselves beneficent and kind." With his suggestion that the self-creative thrust has in technology been followed "too fondly" and "too far," and that bondage has been created under the disguise of good, a profound questioning is introduced. But unlike the ancients, who called for specific delimitations on technics, with the romantics there is no clear outcome other than a critical uneasiness—or a heightened aesthetic sensibility.

Later, in a sonnet on "Steamboats, Viaducts, and Railways" (1835), having observed contradictions between the practical and aesthetic qualities of such artifacts, Wordsworth concludes that

> In spite of all that beauty may disown
> In your harsh features, Nature doth embrace
> Her lawful offspring in Man's art; and Time,
> Pleased with your triumphs o'er his brother Space,
> Accepts from your bold hands the proffered crown
> Of hope, and smiles on you with cheer sublime.

Once again technology, in Enlightenment fashion, is viewed as an extension of nature, and even described in Baconian terms as the triumph of time over space.[50] The "lawful offspring" is nevertheless ugly, full of "harsh features" that beauty disowns. Yet from the "bold hands" of technology temporal change is given the "crown of hope . . . with cheer sublime" that things will work out for the good. In Wordsworth's own commentary on *The Excursion*, the problem "is an ill-regulated and excessive application of powers so admirable in themselves."[51] But it is precisely this ill-regulated and excessive technology that also gives birth to a new kind of admiration, that of the sublime.

With regard to the moral character of technology, ambiguity is even more apparent. Consider, for instance, the arguments of Jean-Jacques Rousseau (1712–78), a man who is, in important respects, the founder of the romantic movement, and whose critique takes shape even before the inauguration of the Industrial Revolution itself, strictly in reaction to ideas expressed by the *philosophes*. In 1750, in a prize-winning "Discourse on the Moral Effects of the Arts and Sciences," critical of the kinds of ideas that would shortly be voiced by D'Alembert's "Preliminary Discourse," Rousseau boldly concludes that "as the conveniences of life increase, as the arts are brought to perfection, and luxury spreads, true courage flags, the virtues disappear."[52] "Money, though it buys everything else," he argues, "cannot buy morals and citizens.[53] "The politicians of the ancient world," he says, "were always talking of morals and virtue; ours speak of nothing but commerce and money."[54] In fact, from Rousseau's point of view, not only have "our minds . . . been corrupted in proportion as the arts and sciences have improved,"[55] but the arts and sciences themselves "owe their birth to our vices."[56] Action, even destructive action, particularly on a grand (or sublime) scale, is preferable to nonaction.[57]

What sounds, at first, like a straightforward return to the moral principles of the ancients, however, is made in the name of quite different ideals. Virtue, for Rousseau, is not the same thing it is for Plato or Aristotle—as is clearly indicated by his praise of Francis Bacon, "perhaps the greatest of philosophers."[58] In agreement with Bacon, Rousseau criticizes "moral philosophy" as an outgrowth of "human pride,"[59] as well as the hiatus between knowledge and power, thought and action, that he finds to be a mark of civilization; instead, he praises those who are able to act decisively in the world, to alter it in their favor, even when these are men whom the Greeks would have considered barbarians. Virtue, for instance, lies with the Scythians who conquered Persia, not the Persians; with the Goths who conquered Rome, not the Romans; with the Franks who conquered the Gauls, or the Saxons who conquered England.[60] In civilized countries, he says, "There are a thousand prizes for fine discourses, and none for good action."[61]

With Bacon, Rousseau argues the need for actions, not words, and approves the initial achievements of the Renaissance in freeing humanity from a barren medieval scholasticism.[62] But unlike Bacon, Rousseau sees that even scientific rationality, through the alienation of affection, can often weaken the determination and commitment needed for decisive action. Thus, in a paradox that will become a hallmark of romanticism, Rousseau turns against technology—but in the name of ideals that are at the heart of technology. He criticizes a particular historical embodiment of technology, but only to advance a project that has become momentarily or partially impotent.

It was in England, however, where the Industrial Revolution found its earliest full-scale manifestation, that this paradoxical critique achieved an initial broad literary expression. Such expression took a realistic turn, rejecting classical patterns in favor of the specific depiction of real situations often in unconventional forms. A poem such as William Blake's "London" (1794) or a novel like Charles Dickens *Hard Times* (1854), in their presentation of the dehumanizing consequences of factory labor, equally well illustrate the force of this approach. Wordsworth, again, may be quoted to extend the issue of the alienation of affections to the social level. In a letter from 1801 he writes:

It appears to me that the most calamitous effect which has followed the measures which have lately been pursued in this country, is a rapid decay of the domestic affections among the lower orders of society. . . . For many years past, the tendency of society, amongst all the nations of Europe, has been to produce it; but recently, by the spreading of manufactures through every part of the country . . . the bonds of domestic feeling . . . have been weakened, and in innumerable instances entirely destroyed. . . . If this is true, . . . no greater curse can befall a land.[63]

Romantic realism is, however, allied with visionary symbolism and, through this, epistemological issues. Consider, for instance, another aspect of Blake's genius, his prophetic poems. John Milton over a century before had in *Paradise Lost* (1667) already identified Satan with the technical activities of mining, smelting, forging, and molding the metals of hell into the city of Pandemonium.[64] Following this lead, Blake, in *Milton* (1804), identifies Satan with the abused powers of technology—and Newtonian science. Satan, "Prince of the Starry Hosts and of the Wheels of Heaven," also has the job of turning "the [textile] Mills day & night."[65] But in the prefatory lyric that opens this apocalyptic epic, Blake rejects the necessity of "these dark Satanic Mills" and cries out

> I will not cease from Mental Fights,
> Nor shall my Sword sleep in my hand
> Till we have built Jerusalem
> In England's green & pleasant Land.

This lyric, "And Did Those Feet in Ancient Time," is set to music and becomes the anthem of the British Socialism. A visionary, imaginative—not to say utopian—socialism is the romantic answer to the romantic critique of the moral limitations of technology. Mary Shelley's *Frankenstein* (1818), in another instance, likewise presents a love-hate relationship with technology in which that which is hated is properly redeemed not by premodern delimitation but by the affective correlate of an expansive imagination—namely, love.

Industrialization, then, undermines affection—that is, feeling and emotion, at both the individual and social levels. And this practical fact readily becomes allied with a more theoretical criticism of the Enlightenment emphasis on reason as the sole or principle cognitive faculty. The Enlightenment argued for the primacy of reason as the only means to advance human freedom from material limitations. According to the romantic reply, not only does such an emphasis on reason not free humanity from material bonds (witness the evils of the Industrial Revolution), but in itself it is (in the words of William Blake) a "mind-forged manacle." The focus on reason is itself a limitation that must be overcome; and through the consequent liberation of imagination the historical condition of technical activity can in turn be altered. In the "classic" epistemological defense and definition of Samuel Taylor Coleridge:

The imagination . . . I consider either as primary, or secondary. The primary imagination I hold to be the living power and prime agent of all human perception, and as a repetition in the finite mind of the eternal act of creation in the infinite I AM. The secondary I consider as an echo of the former, co-existing with the conscious will, yet still as identical with the primary in the kind of its agency, and differing only in degree, and in the mode of its operation. It dissolves, diffuses, dissipates, in order to re-create; or where this process is rendered impossible, yet still, at all events, it struggles to idealize and to unify.[66]

Indeed, it is to this power that Blake also appeals as the source of his social revolution, when he proclaims, "I know of no other Christianity and of no other Gospel than the liberty both of body & mind to exercise the Divine Arts of Imagination, the real & eternal World of which this Vegetable Universe is but a faint shadow, & in which we shall live in our Eternal or Imaginative Bodies when these Vegetable Mortal Bodies are no more."[67]

Finally, with regard to artifacts, the romantic view is again both like and unlike that of the Enlightenment. It is Enlightenment-like in the belief that nature and artifice operate by the same principles. Contra the Enlightenment, however, the romantic view takes nature as the key to artifice rather than artifice as the key to nature. The machine is a diminished form of life, not life a complex machine. Furthermore, nature is no longer perceived primarily in terms of stable forms; the reality of nature is one of process and change. Wordsworth and other English romantics are taken with the "mutability" of nature. Lord Byron, for instance, at the conclusion of *Childe Harold's Pilgrimage* (1818), when he aspires "to mingle with the Universe, and feel / What I can ne'er express" (canto 4, stanza 177), describes nature as the

> . . . glorious mirror, where the Almighty's form
> Glasses itself in tempests; in all time,
> Calm or convulsed—in breeze, or gale, or storm—
> Icing the Pole, or in the torrid clime
> Dark-heaving—boundless, endless, and sublime—
> The image of Eternity. . . .

(canto 4, stanza 183)

Nature, thus reconceptualized, reflects its new character into the world of artifice.

For the Enlightenment, nature and artifice both exhibit at their highest levels of reality various aspects of mechanical order, the interlocking of parts in a mathematical interrelation of the well-drafted lines of a Euclidean geometry. The metaphysical character of such reality is manifest to the senses through a "classical" vision of the beautiful—although there develops an Enlightenment excitement with the great or grandiose and the consequent projecting of art beyond nature that contradicts the models of harmonious stability within nature charateristic of classical antiquity and thus intimates romantic sensibilities. For romanticism, by contrast, the metaphysical reality of both nature and artifice is best denoted not by stable or well-ordered form but by process or change, especially as apprehended by the new aesthetic category of the sublime or the overwhelming and what Byron refers to as "pleasing fear" (canto 4, stanza 184).

As an aesthetic category, the idea of the sublime can be traced back to Longinus (third century A.D.) who departed from classical canons of criticism by praising literature that could provoke "ecstasy." But the concept received little real emphasis until Edmund Burke's *A Philosophical Enquiry into the Origin of Our Ideas of the Sublime and Beautiful* (1757). For Burke, beauty is associated with social order and is represented with harmony and proportion in word and figure; the sublime, by contrast, is concerned with the individual striving and is indicated by magnitude and broken line. "Whatever is fitted in any sort to excite the ideas of pain, and danger, whatever is in any sort terrible, or is conversant about terrible objects, or operates in a manner analogous to terror, is a source of the *sublime*," is Burke's famous definition.[68] Certainly modern technological objects and actions—from Hiroshima to Chernobyl— have tended to become a primary objective correlative of such a sentiment.

Like premodern skepticism and Enlightenment optimism the romantic way of being-with technology can thus be characterized by a pluralism of ideas that constitute a critical uneasiness: (1) the will to technology is a necessary self-creative act, which nevertheless tends to overstep its rightful bounds; (2) technology makes possible a new material freedom but alienates from the decisive strength to exercise it and creates wealth while undermining social affection; (3) scientific knowledge and reason are criticized in the name of imagination; and (4) artifacts are characterized more by process than by structure and invested with a new ambiguity associated with the category of the sublime. The attractive and repulsive interest revealed by the sublime expresses perhaps better than any other the uniqueness of the romantic way of being-with technology.

SUMMARY AND EPILOGUE

As the analysis of romantic being-with technology has especially tended to indicate, the ideas associated with the four aspects of technology as volition, as activity, as knowledge, and as object cannot be completely separated. Theology, ethics, epistemology, and metaphysics are ultimately aspects of a way of being in the world. Acknowledging this limitation, it is nevertheless possible to summarize the three ways of life in relation to technology by means of the matrix displayed in table 1.

At the outset, however, the argument of this essay was indicated

Table 1
Historicophilosophical epochs

Conceptual elements	Ancient skepticism (suspicious of technology)	Enlightenment optimism (promotion of technology)	Romantic uneasiness (ambiguous about technology)
Volition or intention (religious)	Will to technology involves tendency to turn away from God or the gods.	Will to technology is ordained by God or by Nature.	Will to technology is an aspect of creativity—which tends to crowd out other aspects.
Action (ethics and politics)	Personal: Technical affluence undermines individual virtue. Societal: Technical change weakens political stability.	Personal: Technical activities socialize individuals. Societal: Technology creates public wealth.	Personal: Technology engenders freedom but alienates from affective strength to exercise it. Societal: Technology weakens social bonds of affection.
Knowledge (epistemology)	Technical information is not true wisdom.	Technical engagement with the world yields true knowledge (pragmatism).	Imagination and vision are more crucial than technical knowledge.
Objects (metaphysics and aesthetics)	Artifacts are less real than natural objects and thus require external guidance.	Nature and artifice operate by the same mechanical principles.	Artifacts expand the processes of life and reveal the sublime.

to have some relation to Heidegger's early analysis of technology, although it has taken off in a trajectory not wholly consistent with Heidegger's own analysis or intentions. Yet there remains a final affinity worth noting. In Heidegger's existential analysis there is a paradox that the personal that is revealed through the technical is also undermined thereby. The use of tools is with others and in a world of artifacts owned by others, but the others easily become treated as all the same and thus become, as he calls it, a They—mass society.

> In utilizing public means of transport and in making use of information services such as the newspapers every Other [person] is like the next. The Being-with-one-another dissolves one's own *Dasein* [or existence] completely into the kind of Being of "the Others," in such a way, indeed, that the Others, as distinguishable and explicit, vanish more and more.[69]

With regard to the romantic way of being-with technology there is also a paradox. Not only is there a certain ambiguity built into this attitude, but the attitude itself has not been adopted in any wholehearted way by modern culture. Romanticism is, if you will, uneasy with itself. Indeed, this may be in part why romanticism has so far been unable to demonstrate the kind of practical efficacy exhibited by both premodern skepticism and Enlightenment optimism. The paradox of the romantic way of being-with technology is that, despite an intellectual cogency and expressive power, it has yet to take hold as a truly viable way of life. Given almost two centuries of active articulation, this impotence may well point toward inherent weaknesses. Perhaps the truth is that romanticism has been adopted, but that it is precisely its internal ambiguities, its bipolar attempt to steer a middle course between premodern skepticism and Enlightenment optimism, that vitiates its power.

NOTES

1. Martin Heidegger, *Being and Time*, trans. John Macquarrie and Edward Robinson (New York: Harper and Row, 1962), p. 154.

2. One *locus classicus* of such celebration is Sophocles, *Antigone* 332.

3. For an interpretation of the specifically religious dimensions of this negative mythology, see Carl Mitcham, "'The Love of Technology Is the Root of All Evil'," *Epiphany Journal* 8, no. 1. (Fall 1987): 17–28.

4. For Plato, see especially the fifth book of the *Laws* (743d), where agriculture is described as keeping production within proper limits and as helping to focus attention on the care of the soul and the body. Cf. also *Laws*, bk. 8 (842d–e) and bk. 10 (889d).

5. For Aristotle, see especially the *Politics* 1.8–11, and the distinction between two ways of acquiring goods, agriculture and business, the former of which is said to be "by nature" (1258a38), the latter "not by nature" (1258b41). In the *Politics* 6.2, agrarian based democracy is described as both "oldest" and "best" (1318b7–8).

6. Following Aristotle, St. Thomas Aquinas's commentary on the *Politics* terms farming "natural," "necessary," and "praiseworthy" (*Sententia libri Politicorum* 1, lectio 8), and again in *De regimine principum* 2.3, Thomas identifies farming as "better" than commercial activities for providing for material welfare. For Thomas Aquinas, however, farming tends to be spoken of in relation to all manual labor, and in consequence of the doctrine of the Fall takes on a certain ambiguity not found in Aristotle. For instance, in the *Summa theologiae* 2.2, quaestio 187, articulus 3, "Whether Religious Are Bound to Manual Labour," it is argued that all human beings must work with their hands for four reasons: to obtain food (as proof texts Thomas cites Gen. 3.19 and Ps. 128.2), to avoid idleness (Sirach 33.27), to restrain concupiscence by mortifying the body (2 Cor. 6.4–6), and to enable one to give alms (Eph. 4.28). Note that there is a subtle difference between the first two reasons (which cite the Hebrew Scriptures) and the second (which cite the Greek Scriptures). For a relevant interpretation of Thomas Aquinas's thought, which nevertheless fails to recognize the tensions alluded to here, see George H. Speltz, *The Importance of Rural Life according to the Philosophy of St. Thomas Aquinas* (Washington, D.C.: Catholic University of American Press, 1945). Cf. also Philo, "De agricultura," a commentary on Noah as farmer.

7. "Those who labour in the earth are the chosen people of God, if ever he had a chosen people, whose breasts he has made his peculiar deposit for substantial and genuine virtue. . . . Corruption of morals in the mass of cultivators is a phaenomenon of which no age nor nation has furnished an example." Thomas Jefferson, *Notes on the State of Virginia* (1782) (Chapel Hill: University of North Carolina Press, 1955), Query 19, "Manufactures." See also a letter to John Jay, 23 August 1785: "Cultivators of the earth are the most valuable citizens. They are the most vigorous, the most independent, the most virtuous, and they are tied to their country, and wedded to its liberty and interest, by the most lasting bonds. As long, therefore, as they can find employment in this line, I would not convert them into mariners, artisans or anything else."

8. Xenophon, *Memorabilia* (Oxford: Clarendon Press, 1890) 1.1.7, 1.1.9, and Cf. 4.7.10.

9. Cf. also *Memorabilia* 4.7.6–7.

10. Cf. Empedocles, frag. 111.

11. Cicero, *Tusculan Disputations* 5.4.10–11. See also *Academica* 1.4.15.

12. For a development of this argument, see "Philosophy and the History of Technolgoy," *The History and Philosophy of Technology*, ed. George Bugliarello and Dean B. Doner (Urbana: University of Illinois Press, 1979), pp. 163–201.

13. Compare Aristotle, *Physics* 2.1.193a12–17; *Politics* 7.7.1337a2; and *Oeconomica* 1.1.1343a26–b2.

14. Plutarch, "Life of Marcellus," in *Plutarch's Lives*, ed. A. H. Clough (Boston: Little, Brown, and Company, 1859), 2.252–55.

15. On the inadequacy of human knowledge, see the Book of Job, Prov. 1.7, Isa. 44.25, and Col. 2.8. Power over the world, Satan says in the Gospel of Luke, has been given to him (Luke 4.6). The prince of this world, according to the Gospel of John, is to be cast out (John 12.31).

16. In his study of Milton (in "The Lives of the Poets," 1.99–100 [paragraphs

39–41]), Samuel Johnson criticizes a program of education which would concentrate on natural philosophy. "The truth is, that the knowledge of external nature, and the sciences which that knowledge requires or includes, are not the great or the frequent business of the human mind. Whether we provide for action of conversation, . . . the first requisite is the religious and moral knowledge of right and wrong . . . . Physiological learning is of such rare emergence, that one may know another half his life, without being able to estimate his skill in hydrostatics or astronomy; but his moral and prudential character immediately appears. . . . [And] if I have Milton against me, I have Socrates on my side. It was his labour to turn philosophy from the study of nature to speculations upon life; but the innovators whom I oppose . . . seem to think, that we are placed here to watch the growth of plants, or the motions of stars. Socrates was rather of the opinion that what he had to learn was, how to do good, and avoid evil." Cf. also "The Rambler" no. 24 (Saturday, 9 June 1750).

17. Norbert Wiener, "A Scientist Rebels," *Bulletin of the Atomic Scientists* 3, no. 1 (January 1947): 31.

18. John Wesley, *Works* (Grand Rapids, Mich.: Zondervan, n.d. [photomechanical reprint of the edition published by the Wesleyan Conference, London, 1872]), 7: 289.

19. Martin Heidegger, "The Thing," in *Poetry, Language, Thought*, trans. Albert Hofstadter (New York: Harper and Row, 1971), p. 166. See also Heidegger's essay on Rike, "What Are Poets For?" esp. pp. 112–17.

20. Francis Bacon, "Preface," *The Great Instauration*, in *Selected Writings of Francis Bacon*, ed. Hugh G. Dick (New York: Modern Library, 1955).

21. Bacon, *Essays*, no. 24, "Of Innovations," in *Selected Writings of Francis Bacon*, p. 65.

22. Francis Bacon, *Novum Organum*, in *The Works of Francis Bacon* (Stuttgart-Bad Constatt: Friedrich Fromann Verlag Gunther Holzboog, 1963), 1.68. Subsequent references to the *Novum Organum* will appear parenthetically within the text.

23. According to the Talmud, "As God fills the entire universe, so does the soul fill the whole body" (Berakhot 10a). According to the teachings of Jesus, "Love your enemies and pray for those who persecute you, so that you may be sons of your Father who is heaven; for he makes his sun rise on the evil and on the good, and sends rain on the just and on the unjust" (Matt. 5.44–45).

24. For a study of this transformation in the history of ideas, see C. E. Trinkhous, *In Our Image and Likeness: Humanity and Divinity in Italian Humanist Thought*, 2 vols. (London: Constable, 1970).

25. Jean Le Rond D'Alembert, *Preliminary Discourse to the Encyclopedia Of Diderot*, trans. Richard N. Schwab and Walter E. Rex (Indianapolis, Ind.: Bobbs-Merrill, 1963), p. 42.

26. Immanuel Kant, "Idea for a Universal History from a Cosmopolitan Point of View" (1784), Third Thesis, in *On History*, trans. Lewis White Beck (Indianapolis, Ind.: Bobbs-Merrill, 1963), p. 13.

27. Immanuel Kant, "What Is Enlightenment?" (1.784), in *On History*, p. 3.

28. See Aristotle, *Politics* 1268b25–1269a25; and St. Thomas Aquinas, *Summa theologiae* 1–2, quaestio 97, articulus 2.

29. Charles-François de Saint-Lambert, "Luxury," in *Encyclopedia: Selections*, trans. Nelly S. Hoyt and Thomas Cassirer (Indianapolis, Ind.: Bobbs-Merrill, 1965), p. 204 (translation slightly altered).

30. Saint-Lambert, "Luxury," p. 231.

31. David Hume, *Essays* (London: Oxford University Press, 1963), p. 262.

32. Ibid., p. 276.

33. Ibid., p. 277.

34. Ibid., pp. 277–78.

35. Charles Montesquieu, *The Spirit of the Laws*, trans. Thomas Nugent, ed. David Wallace Carrithers (Berkeley: University of California Press, 1977), vol. 1, bk. 20: 1.

36. See Francis Bacon, *The Great Instauration*, "The Plan of the Work," in *Selected Writings*.

37. D'Alembert, *Preliminary Discourse*, p. 75.

38. Ibid., p. 122.

39. Denis Diderot, "Art," in *Encyclopedia*, p. 5.

40. Ibid., p. 4.

41. For some discussion of this contrast, see Nicholas Lobkowicz, *Theory and Practice: History of a Concept from Aristotle to Marx* (Notre Dame, Ind.: University of Notre Dame Press, 1967).

42. On this interesting topic, see K. J. H. Berland, "Bringing Philosophy Down from the Heavens: Socrates and the New Science," *Journal of the History of Ideas* 47, no. 2 (April–June 1986): 299–308, a commentary on Amyas Busche's *Socrates: A Dramatic Poem* (1758). One point Berland does not consider is the extent to which this view of Socrates, which is also found in Aristophanes' *The Clouds* as well as other sources, might be legitimate; see, e.g., Leo Strauss, *Socrates and Aristophanes* (New York: Basic Books, 1966).

43. Alexander Pope, *An Essay on Man*, epistle 1, line 289.

44. Thomas Hobbes, *Leviathan, or the Matter, Forme, & Power of a Common Wealth Ecclesiasticall and Civill* (1651), edited by C. B. Macpherson (New York: Penguin Books, 1968), p. 81.

45. Julien Offray de La Mettrie, *Man a Machine* (*L'Homme Machine*, 1748) (La Salle, Ill.: Open Court, 1912), p. 148.

46. Hobbes, *Leviathan*, "Introduction," p. 23.

47. This is vividly demonstrated by the vicissitudes of development now taking place in countries of the Third World. Geographic advantage, scientific knowledge, imported hardware, political or economic decisions, piecemeal optimism, and envious desire cannot by themselves or even in concert effect industrialization. Despite the ideological rhetoric of Maoist China and Khomeni's Iran, modern technology does not seem to be adopted independently of certain key elements of Western culture. The westernization of Japan confirms the argument from the other side of the divide.

48. For one collection of texts that does begin to point in this direction, see Humphrey Jennings, *Pandaemonium: The Coming of the Machine as Seen by Contemporary Observers, 1660–1886*, ed. Mary-Lou Jennings and Charles Madge (New York: Free Press, 1985). Note that even Leo Marx in a later essay continues to speak of science as the focus of concern; see "Reflections on the Neo-Romantic Critique of Science," *Daedalus* 107, no. 2 (Spring 1978): 61–74. However, in *The Machine in the Garden: Technology and the Pastoral Ideal in America* (New York: Oxford University Press, 1964) he describes one aspect of the romantic critique of technology. Cf. also Wylie Sypher, *Literature and Technology: The Alien Vision* (New York: Random House, 1968).

49. Cf. Friedrich Nietzsche, *The Gay Science* (1882), section 12. For a mundane philosophy of gothic pathos, see Jean-Paul Sartre, *Being and Nothingness* (1943), trans. Hazel Barnes (New York: Washington Square Press, 1968), p. 784, the last sentence of the last chapter of which declares that "Man is a useless passion."

50. See Francis Bacon, "The Masculine Birth of Time," translation included in Benjamin Farrington, *The Philosophy of Francis Bacon* (Chicago: University of Chicago Press, 1966).

51. The note is to *The Excursion*, bk. 8, line 112, at the beginning of a passage describing the industrial transformation of the English landscape as one in which "where not a habitation stood before, / Abodes of men" are now "irregularly massed / Like trees in forests" (lines 122–24) and as a "trumph that proclaims / How much the mild Directress of the plough / Owes to alliance with these new-born arts!" (lines 130–32). "In treating of this subject," Wordsworth writes in his note, "it was impossible not to recollect, with gratitude, the pleasing picture . . . Dyer has given of the influences of manufacturing industry upon the face of this Island. He wrote at a time when machinery was first beginning to be introduced, and his benevolent heart prompted him to auger from it nothing but good." Wordsworth, as much as Sophocles (*Antigone*, 331), is capable of appreciating the benefits of technology. But, he adds, now "Truth has compelled me to dwell upon the baneful effects arising out of an ill-regulated and excessive application of powers so admirable in themselves."

52. Jean-Jacques Rousseau, "A Discourse on the Arts and Sciences," in *The Social Contract and Discourses*, trans. G. D. H. Cole (New York: Dutton, 1950), p. 164.

53. Ibid., p. 162.

54. Ibid., p. 161.

55. Ibid., p. 150.

56. Ibid., pp. 158–59.

57. Cf. in this same regard, Niccolo Machiavelli's use of *virtú* as power in *The Prince* (1512).

58. Rousseau, "A Discourse on the Arts and Science," p. 173. The encyclopedics likewise praise Bacon above all other philosophers.

59. Ibid., p. 158.

60. Ibid., p. 152.

61. Ibid., p. 168.

62. Ibid., p. 146.

63. William Wordsworth, letter to Charles James Fox, 14 January 1801. In this commentary on his presentation of "domestic affections" in the poems "The Brothers" and "Michael," Wordsworth further remarks that "The evil [of the destruction of domestic affections] would be the less to be regretted, if these institutions [of industrialization] were regarded only as palliatives to a disease [in a manner not unlike that associated with ancient skepticism]; but the vanity band pride of their promoters are so subtly interwoven with them, that they are deemed great discoveries and blessings to humanity [as per Enlightenment optimism]."

64. See John Milton, *Paradise Lost*, bk. 1, line 670. Milton also associates Satan's legions with engines and engineering at bk. 1, line 750 and bk. 6, line 553.

65. William Blake, *Milton* (1804), ed. Kay Pankhurst Esson and Roger R. Esson (Boulder, Colo.: Shambhala, 1978), bk. 1, sec. 4, lines 9–10.

66. Samuel Taylor Coleridge, *Biographia Literaria*, ed. George Watson (New York: Dutton, 1956), chap. 13, p. 167.

67. William Blake, *Jerusalem*, pt. 4, "To the Christians," introduction (London: George Allen & Unwin, 1964).

68. Edmund Burke, *A Philosophical Enquiry into the Origin of our Ideas of the Sublime and Beautiful* (1757) (London: R. & J. Dodsley, 1759). pt. 1, sec. 7, first sentence.

69. Heidegger, *Being and Time*, p. 164.

Technological Consciousness and the Modern Understanding of the Good Life

JAMES L. WISER

1

In this examination of modern technology, it may be helpful to differentiate among three distinct levels of meaning that are commonly associated with the term.[1] The most obvious aspect of technology refers to the various types of knowledge, skill, and equipment that are used in the pursuit of essentially pragmatic ends. It is symbolized by the machine, and its impact upon the modern world has been immense. For example, the systematic introduction of machinery into the production process has created a form of industrialized society that is characterized by the following traits: (1) a high density of population; (2) the prolongation of life; (3) urbanization; (4) increases in the per capita amount of material goods; and (5) economic instability.[2] This same process, in turn, has created an international situation in which those nations that have not yet developed an efficient industrial economy continue to suffer because of this condition. On the one hand underdeveloped nations are at a military disadvantage, while on the other they are susceptible to exploitation as industrial nations seek access to both raw materials and foreign markets. As population pressures increase and as the gap between the affluent and the impoverished widens, political tensions are bound to increase.[3]

A second level of meaning associated with the concept of technology refers to those specific forms of organization by which the first-level skills, knowledge, and equipment are administered. Here the dominant reality is that of the bureaucracy. The growth of bureaucracies appears to be intimately linked with the development of large-scale technological production. Consequently, the "bureauc-ratization" of power appears to be a characteristic feature of the industrial society as such, as both its socialist and capitalist forms increasingly exhibit the impersonalism and structural rigidity associated with bureaucratic organizations.[4]

The third level of meaning and the most encompassing refers to those basic assumptions, beliefs, and expectations that give direction to and sustain the operation of both the machine and the bureaucracy. There is, so to speak, a technological mind informed by a fairly explicit set of values that understands technological society as "the mode and condition of its self-realization."[5] As such it is possible to speak of a technological consciousness that is characterized by its particular structure and expressed in specific intellectual and cultural forms. An examination of this technological mind will be the primary task of this study.[6]

The reason to focus our attention upon technological consciousness is given by the very subject matter under consideration. Our concern is with *modern* technology, not technology per se. As tool-making animals, humans have always displayed some form of productive skill. Consequently the organization of those skills, however primitive, is a universal feature of human society. Technology itself is not new. There is no doubt that modern technology is more efficient, more complex, and more extensive than the premodern varieties and that these quantitative differences are important. Yet they are not crucial. What is characteristic of modern technology, what makes it particularly modern, is our attitude toward it.[7]

Today the practice of technology is granted a greater dignity and importance than ever before. Although we may no longer believe that it is an unmixed blessing, technological accomplishment is still appreciated as a high form of human excellence. The good life is commonly defined in material terms, and political activity, consequently, is increasingly concerned simply with the administration of economic affairs.

Traditionally the practice of technology was understood as a legitimate, yet properly restricted, human undertaking. Technology served material needs to supposedly higher concerns. For example, in the writings of Plato and Aristotle, the life of production and commerce was considered to be inferior to both the political life and the life of contemplation. As such, production beyond a certain level was inappropriate and a life devoted exclusively to laboring was considered the least noble of callings.

The development of modern technology, on the other hand, has been characterized by a process of liberation in which the traditional restraints on technological activity have been more or less effectively removed. Such a liberation was possible only because the traditional constraints no longer appeared to make "sense." Lacking only apparent meaning, they also lacked legitimacy. This question of legitimacy, in turn, refers us once again to the importance of understanding the structure of technological consciousness for it is the structure of consciousness that ultimately provides the principles for social and political order.

The structure of the technological mind is no more visible than in the work of Francis Bacon and René Descartes, two of the founders of modern science. Although from one perspective they may appear as opposites—one an empiricist, the other a rationalist—they share in common the central features of the modern scientific project. A comparison of certain themes in their writings with similar themes found in the work of Plato and Aristotle may help to illustrate the fundamental differences between modern and premodern technological consciousness.

2

To compare modern science with its premodern form is, in fact, to follow the recommendations of both Bacon and Descartes. Each understood himself as attempting to break radically from the available intellectual traditions of the time. Descartes, for example, assumed that reality was ultimately simple. Given this, a true knowledge of it should be clear, distinct, and ultimately compelling before any truly open mind. The fact that both the formal and the customary learning of his day exhibited variety, complexity, and disagreement was sufficient proof for Descartes to conclude that the traditional sciences had failed to establish a truly adequate understanding of nature. He resolved, therefore, "to seek no other knowledge than that which I might find within myself, or perhaps in the great book nature."[8]

In the same spirit, but independently of Descartes, Francis Bacon arrived at a similar conclusion. His review of traditional science and of the Greek natural philosophy upon which it was based forced him to conclude that both were incapable of any real development or progress. In particular, the traditional sciences were based on an inadequate logic and lacked thereby a proper method for controlling the idols of the mind. Inasmuch as renewal and reform were impossible, the only alternative was to begin anew. Consequently, in his *Proem of the Great Instauration*, Bacon wrote:

Now that the errors which have hitherto prevailed, and which will prevail for ever, should. . . either by natural force of the understanding or by help of the aids and instruments of logic, one by one, correct themselves, was not a thing to be hoped for, because the primary notions of things. . . are false, confused, and overhastily abstracted from facts; nor are the secondary and subsequent notions less arbitrary and inconstant; whence it follows that the entire fabric of human reason which we employ in the inquisition of nature is badly put together and built up, and like some magnificent structure without any foundation. . . . There was but one course left, and therefore to try the whole thing anew upon a better plan, and to commence a total reconstruction of sciences, arts, and all human knowledge raised upon the proper foundations.[9]

According to Bacon, one of the most important characteristics of the new science was to be found in the purpose for which it was created. Unlike the traditional science, which was essentially contemplative in nature, the new science was to be primarily an activity concerned with the control and domination of nature. Thus Bacon wrote:

For the end which this science of mine proposes is the invention not of arguments but of arts; not of things in accordance with principles but of principles themselves; not of probable reasons, but of designations and directions for works. And as the intention is different so accordingly is the effect: the effect of one being to overcome an opponent in argument, of the other to command nature in action.[10]

Or in the words of Descartes:

For they had satisfied me in that it is possible to reach knowledge that will be of much utility in this life; and that instead of the speculative philosophy now taught in the schools we can find a practical one, by which. . . we can employ these entities for all the purposes for which they are suited, and so make ourselves masters and possessors of nature.[11]

In part, therefore, modern science originated as humankind's effort to impose its design upon the formless matter of nature. Given this intention, even the theoretical sciences are characterized by their fundamentally technological nature.

This argument may be clarified by contrasting the modern understanding with that of both Plato and Aristotle.[12] For them, the distinction between philosophy and science had not yet emerged. Accordingly, the purpose of science was set within the context of the aims of philosophy, and those aims, in turn, were ultimately of a moral nature. The practice of science was intended to achieve those particular virtues and that specific excellence characteristic of philosophical existence in general. In Plato's thought, by turning away from the world of appearances and toward the reality of the ideas and the Good, the philosopher became attuned to the order of Being. On one hand this attunement produced a self-conscious awareness of the structure of reality while at the same time it effected a reordering of the philosopher's soul that required the rule of wisdom and elicited the virtues of justice and prudence.[13]

Without developing the complicated metaphysics that Plato used to support his understanding, it is possible to isolate several features of his thinking that are most relevant for our argument. First, science, for Plato, was understood as an essentially personal or existential achievement. Its purpose was not to gain control of the material world. On the contrary, the philosopher is the one who must initially turn away from material nature in order to nourish his or her soul's erotic desire for that which is:

> So then, won't we make a sensible apology in saying that it is the nature of the real lover of learning to strive for what is; and he does not tarry by each of the many things opined to *be* but goes forward and does not lose the keenness of his passionate love nor cease from it before he grasps the nature itself of each thing which *is* with the part of the soul fit to grasp a thing of that sort; and it is the part akin to it that is fit. And once near it and coupled with what really is, having begotten intelligence and truth, he knows and lives truly, is nourished and so ceases from his labor pains, but not before.[14]

Second, according to Plato, philosophy is a way of life. As such its success is seen in the virtues it sustains and not in the works that it may produce. Properly speaking its activity is an activity upon the soul, a transformation of the human psyche that seeks to attune it according to the principles of its natural measure.

From this perspective, then, the technological domination of material nature appears to be a distraction rather than a true human achievement. The detailed study of mundane reality that is required by technological activity may be seen as an exercise of *logos* that has in some way become detached from the governing vision of philosophical insight. At best it produces the quality of "much learning" (*polymathie*), which is characteristic of the sophist who, according to Plato, competes with the philosopher for the souls of men and women.

One finds a similar attitude in the writings of Aristotle. In his *Metaphysics*, Aristotle distinguished among three types of science.

Science is either theoretical, practical, or productive. The theoretical sciences are concerned with those realities whose principles of being cannot be other than they are. As such, one is interested in pursuing the theoretical sciences not because of their potential for any immediate use but rather to satisfy one's natural intellectual curiosity. Examples of such a science, according to Aristotle, are physics, mathematics, and theology.

Unlike the theoretical sciences, the practical sciences are concerned with realities that are to some degree dependent upon human action. Humankind possesses a certain degree of control over these objects and exercise this control in order to nurture and perfect them. The two practical sciences that Aristotle mentions are ethics and politics; the challenge in *doing* politics is to do it well. It is done well, in turn, if the leaders choose their policies so as to actualize the purpose of political existence—namely, the pursuit of the common good. For Aristotle such a pursuit is intrinsic to the very nature of the political community itself and the prudent leader is the one who seeks to promote its realization.

Like the practical sciences, the productive sciences exercise a certain degree of control over those realities with which they are concerned. That is to say, like the practical sciences, the productive sciences are concerned with knowledge-in-use. Yet the intended form of use in this case is one that is concerned with making rather than with doing. The productive sciences are concerned with fabrication and are judged according to the standard of efficiency. They correspond to what today is referred to as technology.

The important point for our argument is the hierarchical order that Aristotle establishes among these three types of science. Beginning with the highest, he ranks them as follows: theoretical, practical, and productive. The *bios theoretikos* (the life of contemplation), which corresponds most closely to Plato's understanding of philosophy, is the highest form of human excellence. To achieve it is to enjoy the greatest good, which is happiness, and in so doing, to acquire those intellectual virtues necessary for the human soul to function well. As Aristotle wrote in the *Nicomachean Ethics*:

> However such a life would be more than human. A man who would live it would do so not insofar as he is human, but because there is a divine element within him . . . . So if it is true that intelligence is divine in comparison with man, then a life guided by intelligence is divine in comparison with human life.[15]

The second most excellent form of life is the one devoted to politics. According to Aristotle it is in the governing of society that one establishes those habits and expectations that form and promote the various moral virtues. Finally, Aristotle considers the life of fabrication. As the lowest form of activity, its value is ultimately derived from those essential contributions made in support of the other types. In effect, labor provides those necessary prerequisites without which life—and, therefore, the good life—is impossible.[16] For Aristotle, then, technology has a necessary and therefore a legitimate role; but as a subsidiary activity, it is by definition incapable of providing a focal purpose for human life.

3

A second important feature of the modern scientific revolution is found in modernity's new understanding of nature. Rather than the hierarchically ordered cosmos of the ancients, nature appears to the technological consciousness as simply matter-in-motion. This understanding is described by Alfred North Whitehead as follows:

> There persists, however, throughout the modern period the fixed scientific cosmology which presupposes the ultimate fact of an irreducible brute matter or material spread throughout space in a flux of configurations. In itself such a material is senseless, valueless, purposeless. It just does what it does do, following a fixed routine imposed by external relations which do not spring from the nature of its being.[17]

Again, the implications of such a view can best be illustrated by a comparison of the thought of Bacon and Descartes with that of Plato and Aristotle.

Without going into the details of Plato's cosmology,[18] it is possible to advance our argument simply by referring to his *Phaedo*. Here Plato has Socrates express his disappointment with the reductionism implicit in the cosmogonies of such *physiologoi* as Thales, Democritus, and Anaxagoras. For Socrates, this reductionism is based on an initial confusion concerning causation. According to Socrates, the Ionians are "unable to distinguish between the cause of a thing and the condition without which it could not be a cause. It is this latter, as it seems to me, that most people, groping in the dark, call a cause—attaching to it a name which it has no right."[19] The distinction here is between the material conditions of reality and its rational causes. According to Socrates something is said to exist because it participates in a reality peculiar to its appropriate universal. For example: "whatever else is beautiful apart from absolute beauty is beautiful because it partakes of that absolute beauty and for no other reason."[20]

This understanding is not fully developed by Plato until *The Republic*. There the universal absolutes are given ontological status as ideas. What is important here, however, is his argument that the ideas themselves are caused by and intelligible because of the Good (*agathon*). In this sense, the Good is the cause of all that truly is. As Socrates says in the *Phaedo*:

> Somehow it seemed right that mind should be the cause of everything and I reflected that if this is so, mind producing order sets everything in order and arranges each individual thing in the way it is best for it. Therefore if anyone wished to discover the reason why any given thing came or ceased or continued to be, he must find out how it was best for that thing to be, or to act or be acted upon in any other way. On this view there was only one thing for a man to consider, with regard both to himself and anything else, namely the best and highest good, although this would necessarily imply knowing what is less good, since both were covered by the same knowledge.[21]

Given such an understanding, nature appears as a meaningfully structured whole and is not the result of random forces. As an architectonic whole, it exhibits design and purpose. As such, it has a dignity and stability that provides both a guide to and a limitation upon man's technological adventures.

Although Aristotle took a certain pride in distinguishing his theory of causation from that which Plato has Socrates develop, the resulting view of nature is strikingly similar. Both men understood nature as a purposeful, structured order. Aristotle speaks of four causes: material, efficient, formal, and final. The formal cause refers to the specific determination given to the material at hand. As the governing design that orders the material, the formal cause specifies the particular essence of the reality involved. The final cause is the purpose for which something exists. As the perfected nature of a specific reality, it operates by attraction and establishes the goal against which particular achievements can be measured. In effect, then, nature appears for Aristotle as a drama. Summarizing this view, Whitehead writes: "It thus conceived nature as articulated in the way of a work of dramatic art, for the exemplification of general ideas converging to an end. Nature was differentiated so as to provide its proper end for each thing."[22] An important consequence of this view touches upon the question of technology. If nature is a drama, then man is a spectator. And as a spectator he is primarily

concerned with observing and judging. Consequently control, manipulation, and domination are not seen as totally appropriate human
activities given such a dramatic view of nature. Like Plato's, Aristotle's theory of causation implies a definite limitation upon man's
technological activities.

This particular limitation is removed if the concepts of formal and
final causation are discarded as inappropriate. Historically this is
precisely what happened with the new science of the seventeenth
century. Nature was reduced to matter-in-motion and, as such,
could be adequately explained solely in terms of material and
efficient causation. Such a reduction was implicit in Descartes's suggestion that nature is simply *res extensa* and, as such, devoid of any
intentionality. Intentionality is an attribute of consciousness, and
consciousness, for Descartes, is removed from nature and located in
the abstracted *res cogitans*. A nature that is pure extension can be
explained geometrically and from this perspective the introduction
of such notions as essence or purpose can only serve as a source of
confusion.

Similarly, Bacon explicitly rejects the need for the use of formal
or final causes in any scientific explanation. Although he retains the
term formal cause, he redefines the concept to mean nothing more
than a general law used to describe individual acts. The concept of
final causation, on the other hand, is totally rejected. According to
Bacon the notion of purpose is appropriate only in a description of
human action. Those who attempt to use it in an analysis of nature
therefore are only projecting the category where it does not in fact
belong. Such an act of imposition is typical of those errors that
Bacon describes in his analysis of the idols of the tribe.

If nature is without design or purpose, if it is simply matter-in-
motion, then it is obvious that it contains no "grounded" standard
by which to limit human works. Lacking a formal cause, nature has
no preestablished design that can either be affirmed or imitated.
Lacking a final cause, it is unbounded and thus presents itself as an
indifferent field that is to be ordered by the activities of human
beings. As Hans Jonas writes:

> Nature is not a place where one can look for ends. Efficient causes know
> no preference of outcomes; the complete absence of final causes means
> that nature is indifferent to distinctions in value. It cannot be thwarted
> because it has nothing to achieve. It only proceeds—and its process is
> blind.[23]

Given such an understanding it quickly becomes apparent that the cosmology of modern science implies a specific understanding of humanity's proper role in the universe. Humankind becomes, in effect, the purpose giver in a purposeless world. As a consequence, technology becomes the essential means for human self-actualization and, in so doing, replaces the Greek *bios theoretikos* as the ultimate standard of human excellence.

4

The fact that for modern technological consciousness the life of fabrication has replaced the life of contemplation as the highest form of human excellence has created a significant challenge for contemporary political theory. It faces a situation in which its traditional concepts are becoming increasingly irrelevant. As this situation develops, its ability to analyze and advocate the principles of an acceptable political order becomes increasingly questionable.

Classical political theorists sought to understand the nature of the "good society." Although they may have disagreed on several particulars, they understood in general that the good society and the *bios theoretikos* were intrinsically related. As Eric Voegelin writes:

> According to the classical concept the "good society" is one which: (1) is large enough and wealthy enough to make the life of reason possible, at least for the minority capable of putting this human potentiality to work; (2) is organized in such a way that the life of reason becomes a social force in a society's culture, including its political affairs.[24]

As we have seen the concept of the *bios theoretikos* (life of reason) presupposed a specific understanding of both nature and humanity's proper place within it. As a principled and ordered whole, nature offered a pattern whose structure and design could serve as a source for the principles of human and social order. The life of reason was that form of human existence that was informed by such principles and thus attuned to the measure of Being itself. Its rational substance was derived from its participation in the substantive reason of nature. As such, it was a way of life that was understood as being grounded in the very order of reality itself. Properly speaking then, the life of reason was neither a "value" nor a preferred life style; it was rather the human good as established by the principles of nature in general and of human nature in particular.

With its denial of a hierarchy in nature, on the other hand, mod-

70 JAMES L. WISER

ern natural science asserts a one-leveled ontology. Such an ontology, in turn denies the distinction between that which is naturally
higher and that which is naturally lower. As Marjorie Grene explains:

> First, be it noted, it [the one-leveled universe] is a universe constructed
> on the foundation of the contrast between the *natural* and the artificial.
> There can be no "higher and "lower" in nature. Yet in human life, in
> what we call culture, in language, custom, institutions, we find nature
> transformed by man. . . . To many, these products of human activity,
> laws, theories, works of art, have seemed higher realities, or the expres-
> sions of higher realities to which we owe allegiance. In a one-leveled
> world, they can be interpreted only, in contrast to what "really" is, as
> artifacts, as what we have *made* in contrast to what naturally exists.[25]

In a one-leveled ontology there is no substantive ground for the
life of reason as traditionally understood. Being an artifact, the life
of reason appears as simply one value preference among others. Its
worth is derived only from the fact that it is preferred by some individuals. These preferences, however, have no basis in nature and, as
such, cannot serve as an ontological measure by which to assess the
validity of the various opinions about the good life. With the disappearance of the traditional understanding of the life of reason as a
substantive good, it has become increasingly difficult for political
theorists to articulate a compelling definition of the common good.
This dilemma had become apparent as early as the seventeenth century and it is clearly expressed in the political writings of Thomas
Hobbes.

Hobbes's writings in general are characterized by his self-
conscious attempt to appropriate the truths of modern science.
Many of his specific teachings reflect this effort—for example, his
reduction of both natural and psychic phenomena to matter-in-
motion, his substitution of a *summun malum* for the traditional *summum bonum*, and his understanding of the relationship of power of
its ends. Yet of more interest for our argument is the general spirit
that pervades all of his work. According to Hobbes, nature or that
which really *is*, stands in sharp contrast to the cultural and artificial.
For him such concepts as "the good," the beautiful," and "the just"
do not refer to any preexisting realities that are in some sense rooted
in nature; rather they are but names created by the consent of individuals and hence are legitimate only to the extent that they are
accepted and enforced. Given such an understanding, then, the life
of contemplation becomes just one life style among others. It results
from a particular value preference and, as such, is rooted in the

human will and not in the structure of an essentially indifferent nature. By relativizing the life of contemplation, Hobbes removes the traditional source of political order. His analysis, however, goes further. The life of contemplation is a cultural creation; thus it is in some sense "unreal." A true appreciation of its meaning, therefore, requires that we penetrate to an analysis of its real but hidden foundations. In effect, the *bios theoretikos* has become for Hobbes an example of ideological consciousness. In particular he presents the life of reason as an extreme expression of human self-love and of the lust for power. Consequently, rather than viewing passion as a source for the corruption of the spirit, Hobbes sees it instead as the very foundation of the spiritual life.[26] By this analysis Hobbes has succeeded in completely reversing the traditional classical and Christian interpretation of human order.

This examination of the *bios theoretikos* and of Hobbes's reinterpretation of its significance is intended only as a single illustration of a general situation facing contemporary political theory. It is not surprising that the concepts used in serious political analysis are necessarily based upon specific metaphysical assumptions. As these assumptions are discarded, the corresponding political categories become increasingly meaningless. As we have seen, for traditional political science, the social order of the "good society" presupposes the existential order of the *"bios theoretikos."* This in turn was grounded in a conception of nature that differentiated between a higher and lower order.

As this differentiation has become increasingly difficult to sustain, so too has the traditional concept of the common good lost its substantive grounding. With this loss, the standards of political order have become increasingly procedural in character. Whether a political society that is committed to a procedural rather than to a substantive understanding of order can satisfy the human need for a meaningful social existence is perhaps the central question facing politics in the technological age.

NOTES

1. I am following here the argument of Leo Marx, "Technology and the Study of Man," *The Key Reporter* 39 (Spring 1974): 2–4 and 8.

2. These features and others are discussed by Milford Q. Sibley, *Nature and Civilization: Some Implications for Politics* (Itasca, Ill.: F. E. Peacock, 1977), pp. 201–50.

3. See Robert L. Heilbroner, *An Inquiry into the Human Prospect*, 2d ed. (New York: W. W. Norton, 1980).

4. See Peter Berger, et al., *The Homeless Mind: Modernization and Consciousness* (New York: Random House, 1974).

5. The phrase is Eric Voegelin's from his *The New Science of Politics* (Chicago: University of Chicago Press, 1952), p. 27. He uses it to describe the fact that every form of society develops and is consequently ordered by its own self-interpretation. Society, accordingly, is not simply a fact or an event in the natural world. This point is essential to our argument. We do not really understand modern technological society if we limit our investigation to a study of its external features. Rather we must attempt to appreciate those engendering experiences that are represented within society's institutions.

6. A number of studies are available for guiding such a project. Among them are Herbert Marcuse, *One-Dimensional Man* (Boston: Beacon Press, 1964); Hannah Arendt, *The Human Condition* (Chicago: University of Chicago Press, 1958); Martin Heidegger, *The Question Concerning Technology and Other Essays*, trans. William Lovitt (New York: Harper and Row, 1977); Karl Jaspers, *Man in the Modern Age*, trans. Eden Paul and Cedar Paul (Garden City, N. Y.: Doubleday, 1957); Nikolai Berdjajew, *Der Mensch in der technischen Zivilization* (Vienna: Amandus, 1948); Victor C. Ferkiss, *Technological Man: Myth and Reality* (New York: Mentor, 1969); and Max Horkheimer, *The Eclipse of Reason* (New York: Seabury Press, 1974).

7. "The best principle of delimiting periods in technical evolution is to my judgment furnished by the relation between man and technology, in other words by the conception which man in the course of history held not of this or that particular technology but of the technical function as such." Jose Ortega y Gasset, "Man the Technician" in his *History as a System*, trans. Helene Weyl (New York: W. W. Norton, 1962), p. 141.

8. René Descartes, *Discourse on Method*, trans. Laurence J. Lafleur (Indianapolis: Bobbs-Merrill, 1950), p. 6

9. Francis Bacon, *The New Organon* (Indianapolis, Ind.: Bobbs-Merrill, 1960), pp. 3–4.

10. Bacon, *New Organon*.

11. Descartes, *Discourse on Method*.

12. Although Plato and Aristotle represent only one of the two major Greek scientific traditions—the other being the Ionian natural philosophers—their understanding of the purpose of science is nonetheless representative of Greek thought as a whole. See F. M. Cornford, "Greek Natural Philosophy and Modern Science," in *The Unwritten Philosophy* (New York: Cambridge University Press, 1967), pp. 81–94.

13. The above sketch is a highly compacted account of Plato's "Cave Allegory" in the seventh book of his *Republic*.

14. Plato, *The Republic*, trans. Allan Bloom (New York: Basic Books, 1968), 4901–b.

15. Aristotle, *Nichomachean Ethics*, trans. Martin Ostwald (Indianapolis, Ind.: Bobbs-Merrill, 1962), X, 7, 1177 B 25–30.

16. Hans Jonas, *The Phenomenon of Life: Toward a Philosophical Biology* (New York: Dell, 1966), p. 198.

17. Alfred North Whitehead, *Science and the Modern World* (New York: Mentor, 1948), p. 23.

18. See Gregory Vlastos, *Plato's Universe* (Seattle: University of Washington Press, 1975)

19. Plato, *Phaedo*, trans. Hugh Tredennick in *The Collected Dialogues of Plato*, ed., Edith Hamilton and Huntington Cairns (Princeton: Princeton University Press, 1963), 93a–e.

20. Plato, *Phaedo*, 100e.

21. Plato, *Phaedo*, 97c.

22. Whitehead, *Science*, p. 15.

23. Hans Jonas, "The Scientific and Technological Revolutions," *Philosophy Today* 15 (Summer 1971): 92.

24. Eric Voegelin, "Industrial Society in Search of Reason," in *World Technology and Human Destiny*, ed. Raymond Aron (Ann Arbor: University of Michigan Press, 1963), p. 38.

25. Marjorie Grene, "Hobbes and the Modern Mind," in her *Philosophy in and out of Europe* (Berkeley: University of California Press, 1976), pp. 155–66.

26. For an interpretation of Hobbes that emphasizes this aspect of his thought, see Peter J. Opitz, "Thomas Hobbes," in *Zwischen Revolution und Restauration*, ed. Eric Voegelin (Munich: List, 1968), pp. 47–81.

Technology and the Denial of Mystery: The Sacrilization of the Familiar

DAVID LOVEKIN

I begin with the suggestion that the nineteenth- and twentieth-century fascination with popular mystery novels, short stories, and science fiction is an important clue in fathoming the humiliation of Mystery in our age. I claim, following the insights of contemporary French thinker Jacques Ellul, that the only mystery we are able to respect is that provisional and transient unknown that is open to scientific and technological inspection, an unknown completely compatible with *the familiar*, open to those things and processes that are the results of scientific ratiocination. The scientific or technical unknown is a nontranscendent mystery, which may appear as an outgrowth of culture. The scientific or technical unknown is that which is available to reason. A transcendent sense of mystery, however, is not available to reason but is a function of the imagination. It is this sense of mystery, discussed most profoundly by the eighteenth-century philosopher Giambattista Vico, that makes culture possible. This sense of mystery is tied to what Vico understood as *fantasia*, or imagination, which is opposed to reason, or *ragione*.

According to Ellul, *la technique* or technology is a kind of mentality, a focus of consciousness, located only in the *hic et nunc*, the here and now. *La technique* is a mentality of method, a search for the "one best way," conceived of absolutely and according to a mathematics-like logic. This mentality, all too often identified with machines, is a form of intentionality that makes the machine possible. The machine is merely a symptom of technology. Following Ellul, I wish to urge that the current concern with mystery and science fiction is a further manifestation of *la technique*, the machine in one of its many suits. I challenge the notion of detective stories and science fiction as "imaginative literature."

Ellul understands that *la technique* becomes a clear direction

74

of thought, society, and culture after 1750, at which point Cartesianism—previously a philosophy—is everywhere applied.[1] The true is reduced to the real, to an image before an attentive subject, and embraced in a methodology seeking concepts that are noncontradictory. Descartes believes that the true is clear and distinct and available to a fourfold method where one should: (1) begin with what is clear and distinct; (2) reduce all complexities to their simplest parts; (3) reassemble the analyzed parts to constituent wholes; and (4) engage in a course of constant enumeration.[2] For Descartes, this methodology is the business of the scientist and the philosopher. However, Ellul believes that after 1750 this business broadens and becomes the single motivating force entering all domains: politics, economics, religion, language, and finally art. At this point, technology becomes the sacred, the ultimate mystery that is worthy of respect. Ironically, this is no mystery.

For Ellul, mystery is that which cannot be spelled out in noncontradictory terms; mystery is that which transcends and gives meaning to the hear and the now. Mystery is exhausted in the image and is only preserved in the word, the metaphor, that reveals without denigrating the fullness of a reality that surrounds, a reality that traditionally directed art, religion, and philosophy before the true was confined to the clear and distinct. Contemporary art, then, functions only as a distraction or as a restatement of the concerns of technique; it is Cartesianism applied.

Eighteenth-century philosopher and professor of Latin Eloquence, Giambattista Vico anticipates Ellul's concern. In his 1708 inaugural address to the University of Naples, *De nostri temporis studiorum ratione* (*On the Study Methods of Our Time*), Vico warns of the effects of Cartesianism on the imagination (*fantasia*).[3] Reason and the disciplines of analysis, while valuable, may dry up the imagination and weaken the powers necessary for creating imaginative universals—the metaphors that poets create and that provide a basis for a community's sensibilities.[4] The poets of antiquity named the thundering sky Jove and thereby formed the first communities as well as the first languages. For Vico, mystery plays a great part in this process. In the face of what it does not understand, the imaginative mind makes and forms the unknown out of itself, providing a new shape for experience. Out of this new shape, the fundamentals of culture are born. Reason may understand what the imagination creates, but reason does not create out of this understanding. Reason requires the work of the imagination, without which reason only reduplicates itself and repeats the familiar.

It is my purpose in this essay to connect Vico and Ellul on the matter of mystery and to locate mystery in the modern age in popular literature—in the detective story and in science fiction and fantasy. I do not offer a definitive perspective on these genres. It is not my concern to fine-tune all the distinctions available for the analysis of mystery and science fiction. I wish to identify the levels of technique and Cartesianism present in this denuded art and to show that the modern mystery is simply another version of what has become the *technologically familiar*. I suggest that science fiction and the detective story locate the imagination in its absence.

1

It may be quickly and easily noticed that much prime-time television programming is devoted to problem solving. The endless drift of physicians, lawyers, police, and detectives all take up problems soluble at a single sitting. Difficulties are dispatched nicely between commercial announcements and between weekly episodes. Needed by problem solvers are patience, keen observation, and, most important, proper methodology. In the typical crime drama, we begin with the preparation of the crime, say the committing of a murder, the more complicated the better. The murderer and the victim come together within the first ten to fifteen minutes of the program. The remainder of the show features attempts to determine how the crime was committed, the knowledge of which the audience already possesses. Such programming is not concerned with the motivation of the crime, circumspection upon the reality and meaning of crime itself apart from sociological or psychological dimensions (technical dimensions), or more transcendent considerations of justice and morality. The mystery becomes a matter of uncovering what the audience already knows. All of this is significant, for the same formula is worked through in science fiction and in other forms of popular entertainments.

Edgar Allan Poe, founder of the detective story, devised this formula in *The Purloined Letter*, published in 1845. A letter is stolen. The thief is known. The whereabouts of the letter is for the reader and Monsieur Dupin to find out. Poe's work inspired an entire generation of French literature beginning with Baudelaire. Baudelaire writes, approvingly, in Cartesian fashion: "Poe is always correct. It is a very remarkable fact that a man with such a bold and roving imagination should be at the same time so fond of rules and capable of careful analyses and patient research."[5]

Poe is likely a manifestation of the Cartesian mentality, the paradigm of *la technique*. He devised a theory of literary criticism and an aesthetic based on scientific and technological principles. A work should be brief enough to allow a reading at one sitting and should permit the intrusion of worldly affairs without allowing distractions to affect the all important unity and the effect of the work. In "The Philosophy of Composition," published in 1846, Poe states:

> Within this limit, the extent of a poem may be made to bear mathematical relation to its merit—in other words, to the excitement or elevation—again in other words, to the degree of the true poetical effect which it is capable of inducing; for it is clear that the brevity must be in direct ratio to the intensity of the intended effect—this, with one proviso—that a certain degree of duration is absolutely requisite for the production of any effect at all.[6]

The artist is to be concerned with an effect produced out of a mathematics-like consideration and with a brevity that would not tax the reader. Poe determined that 108 lines was a suitable poetic length that would not exceed the reader's energy and attention, which he estimated to last one half hour. Literary creation, he concluded, had to begin from the point of a *dénouement*. Consciously contrived, the literary work, poem or story, must exhibit an "untying," a movement toward a solution that is, at any point, apparent. Most artists, Poe thought, would rather have the public see the act of creation as a Dionysic frenzy and not examine "the wheels and pinions—the tackle for scene-shifting—the step-ladders and demontraps—the cock feathers, the red paint, and the black patches, which, in ninety-nine cases out of the hundred, constitute the properties of the literary *histrio*."[7] It is unimportant whether Poe indeed worked in this manner or whether he was adopting a pose; what is important is that he was part of an important direction to which all artistic activity would have to respond. With Poe, the notion of "the artist as critic" had arrived.

Most exemplary of Poe's *ratiocinative tales* is "The Murders in the Rue Morgue." A manifesto for detective tales is found in the following line: "It will be found, in fact, that the ingenious are always fanciful, and the *truly* imaginative never otherwise than analytic."[8] Conclusions are deduced and induced, and the solution to the murders at Rue Morgue appears clearly and distinctly. Typical of this process, and exemplary of the seventeenth- and eighteenth-century fascination for springs and devices, is the following:

> Yet the sashes *were* fastened. They *must*, then, have the power of fasten-

ing themselves. There was no escape from this conclusion. I stepped to the unobstructed casement, withdrew the nail with some difficulty, and attempted to raise the sash. It resisted all my efforts, as I had anticipated. A concealed spring must, I now knew, exist, and this corroboration of my idea convinced me that my premises, at least, were correct, however mysterious still appeared the circumstances attending the nails. A careful search soon brought to light the hidden spring. I pressed it, and, satisfied with the discovery, forebore to upraise the sash.[9]

Here we have all the excitement and drama of working a geometrical theorem, of composing an article for *Popular Mechanics*, and of approaching a problem from the mentality of *la technique*. Throughout the story, Poe, the tackle hoister and scene shifter, is clearly at work, transforming the mundane into the mysterious.

Throughout his literary work, Poe toyed with the idea that the purpose of art, in contradistinction to the direction of science, which seeks truth, was to seek the beautiful and the sublime, "the desire of the moth for the star."[10] In "The Poetic Principle," published in 1850, Poe concludes that "the demands of truth are severe."[11] In presenting truth, the writer must write in precise, clear, and terse language. The language of poetry is more ornate, more replete with images and metaphors, which concerned Poe: "he must be theory-mad beyond redemption who, in spite of these differences, shall still persist in attempting to reconcile the obstinate oils and waters of Poetry and Truth."[12] Poe separates beauty, truth, and goodness— what Plato strove to join—largely for considerations of effect.

Instead of systematically and philosophically distinguishing between reason, passion, and moral empathy, he entertains discussion of cause and effect. The concerns of poetry are beauty, which, as stated previously, are best served in direct and severe language that leads the intellect. In poetry, the elevation of the soul in a flight for beauty is matter at hand. Poe writes:

When, indeed, men speak of Beauty, they mean, precisely, not a quality, as is supposed, but an effect—they refer, in short, just to that intense and pure elevation of the *soul*—not of intellect, or of heart—upon which I have commented, and which is experienced in consequence of contemplating "the beautiful." Now I designate Beauty as the province of the poem, merely because it is an obvious rule of Art that effects should be made to spring from direct causes—that objects should be attained through means best adapted for their attainment,—no one as yet having been weak enough to deny that the peculiar elevation alluded to is *most readily* attained in this poem.[13]

Although Poe warns of mixing concerns of truth and beauty, in his discussion of poetic theory he writes of methods, of causes and effects, and of calculations. He separates truth, beauty, and goodness in a fashion that has come to be regarded as modern. And he dwells upon style in the most quantitative and material form. He discusses "beauty" in terms of *tone* and *theme*. Melancholy, he concludes, is the most legitimate poetical tone; and almost of necessity, he realizes that the ultimate poetic subject would be the death of a beautiful woman, the subject of *The Raven* and the locus for his discussion of poetics.[14] Then, in further consideration of artistic effects, Poe happens upon the matter of a refrain. He writes:

> The universality of its [the refrain's] employment sufficed to assure me of its intrinsic value, and spared me the necessity of submitting it to analysis. I considered it, however, with regard to its susceptibility of improvement, and soon saw it to be in a primitive condition. As commonly used, the *refrain*, or burden, not only is limited to lyric verse, but depends for its impression upon the force of monotone—both in sound and thought. The pleasure is deduced solely from the sense of identity—of repetition. I resolved to diversify, and so heighten, the effect, by adhering, in general, to the monotone of sound, while I continually varied that of thought: that is to say, I determined to produce continuously novel effects, by the variation *of the application* of the *refrain*—the refrain itself remaining, for the most part, unvaried.[15]

The refrain *nevermore* arrives out of a kind of logical necessity.

Poe appears to be working from the logic of the cliché and the logic of the concept, where the identical and the repeatable have inherent value. Variation is employed as a method, as a law. The poetic eye, however, must not stray from effect.

In matters of effect, Poe castigated Hawthorne for his didactic use of allegory, arguing that one should never stray too far from common sense, which Poe understands in a clearly empirical way. He writes: "One thing is clear, that if allegory ever establishes a fact, it is by dint of overturning a fiction."[16] Poe thought that, in principle, an image should never stray too far from fact. Poe was clearly motivated by complex concerns.

Critic Edward H. Davidson suggests that Poe was attempting to construct a theory of the imagination and intuition that cut between the problem of knowing as a form of fact gathering and of knowing as logical elaboration. Davidson feels this was Poe's goal in his last work *Eureka*. He regards this work as Poe's "scientific religion" in which the power of image making was seen as a creative faculty, distinct from scientific conceptualization, while at the same time having some affinity.[17] Davidson notes Poe's interest in Coleridge's theory of the imagination in this regard. Daniel Hoffman agrees, but in less generous terms: "The philosophical breadth of Coleridge underlies Poe's acute narrowness as the pyramid on the Great Seal of the United States at its summit supports one assured and unblinking eye."[18] As I interpret this, Poe's gaze ultimately may be the single-minded stare of *la technique*. While attempting praise, French critic Remy De Gourmont, in 1904, writes: "Poe is much more representative than Emerson or Walt Whitman. His mind has some practical tendencies. Deprived of literature, he would have been an amazing business man, or promoter of the first order."[19]

It is fascinating to think of this poet turned journalist, the author of "The Imp of the Perverse," as struggling with an epistemological principle when he speaks of that "instinct" that runs counter to all reason and rule, all logic and order. This instinct, Poe suggests, is one not available to phrenology, which greatly interested him. He writes:

> We stand upon the brink of a precipice. We peer into the abyss—we grow sick and dizzy. Our first impulse is to shrink from the danger. Unaccountably we remain. By slow degrees our sickness and dizziness and horror become merged in a cloud of unnamable feeling. By gradation, still more imperceptible, this cloud assumes shape, as did the vapor from the bottle out of which arose the genius in the Arabian Nights.[20]

It is fascinating to understand this genius, this genie, this demon, as the imagination rising in the face of true Mystery, an abyss that yawns beneath our most rational and solid footings; it is fascinating to understand this Mystery of Poe's as an essential form of the Other that provides a driving force in the literature of all cultures; but then we shall never know, in this work, whether Poe was striving for *effect*.

2

Mystery, as I understand it, is not simply that which we do not know. Instead, it is the Wholly Other of what we do not know. It is not any one particular thing but is that which totally surrounds. I will not attempt a clear and distinct definition: I cannot and I could not. This Other is the stuff of the prerational world: the experience of the Sublime for Kant in the *Critique of Judgment*; the all-encompassing *Mysterium Tremendum* that so engrosses Rudolf Otto in his 1917 landmark study, *The Idea of the Holy*. It is, in my view, the major concern for the humanities, a concern that distinguishes them from the sciences.

For Kant, sublime objects, apprehended say in "bold, overhanging, and as it were threatening rocks; clouds piled up in the sky, moving with lightning flashes and thunder peals; volcanoes in all their violence of destruction. . ." stimulate the imagination and "raise the energies of the soul above their accustomed height and discover in us a faculty of resistance of quite different kind, which gives us courage to measure ourselves against the apparent almightiness of nature."[21] The Sublime marks the appearance of supreme opposition that is not merely given to the senses by nature, by reflection, or by cultural convention, although judgment on the Sublime "needs culture (more than judgment upon the beautiful)."[22] This sense of the Sublime, related to the imagination, arises out of an awareness of opposition between the familiar and the all-encompassing. It arises, as well, in opposition to sense experience and to reason, although Kant might not follow me in this. The relation between the Sublime, the imagination, and reason is less than clear. It is important, though, that the imagination is understood to have its own logic, that it is to a degree independent of conceptual understanding, and that it is felt as an element in the apprehension of the Sublime and its appearance in nature as a dynamic force.

From another direction, Rodolf Otto understands the experience of ultimate Mystery as the ground for religious experience. This experience, unlike any other, is not grounded in the categories of reason, although it is an experience with a decidedly a priori character. The experience is rooted in the feeling of a presence of Wholly Other—a fascinating yet terrifying *Mysterium Tremendum*, a *numinous* domain surrounding the *numen* that is of a logically different order from any *phenomenon*. Like the Sublime, it may be at once in nature and yet beyond. Otto states: "So the idea of the sublime is closely similar to that of the numinous and is well adapted to excite it and to be excited by it, while each tends to pass over into the other."[23] The slippery relation between the Sublime and the numinous, both importantly preaesthetic and prereligious, remains problematic. Otto concludes: "it is probable that the feeling of the sublime is itself first aroused and disengaged by the precedent religious feeling—not from itself, but from the rational spirit of man and its *a priori* capacity."[24] The relation between reason and religion here is troublesome, and problematic for Kant as well. Otto, like Kant, cannot resist tying the transcendent to a rational spirit. Poe, in his search for a calculus for creativity, inherits this tradition, which I understand to be essentially Cartesian.

To move beyond Descartes it is essential to turn to the eigh-

teenth-century Neapolitan genius of the imagination, Giambattista
Vico. In his masterwork, the *Scienza Nuova*, Vico founds his "new
science," on the original formative powers of the imagination, what
he terms *fantasia*.[25] The imagination forms universals, *universali
fantastici*, in a grand synthesis of image, metaphor, sensuous experience, language, and institution. Language and history are made at
once with *fantasia*. The first community act is a poetic act.

In Vico's view, reason is independent from and ancillary to
imagination. Reason may analyze what the imagination accom-
plishes, inasmuch as it works in relation to abstraction. Universals—
universali intelligibili—are formed by reason as generalities appear
through a process of locating identities and similarities between
particular things.[26] Similarities give rise to the notion of a class—
Aristotle's theory of definition can be so understood. Defining characteristics express similar and finally necessary qualities. Accidental
qualities are qualities an object need not have. Poe's "Philosophy of
Composition" reveals the workings of reason, as he analyzes *The
Raven* from the standpoint of a fait accompli. When Poe says that all
beautiful poems express a unified effect with a well-controlled refrain, having as a major theme the death of a beautiful woman, he is
saying that *The Ravan* is so structured, and little more. Perhaps *The
Raven* was produced in this manner, as well, although the poem,
then, would have little to do with what Vico would understand as
poetry.

Vico associates the intelligible universal with Cartesian thought,
which can take apart what has already been formed. Ships may be
analyzed according to geometrical principles. A mast and its rigging
form right triangles with measurable hypotenuses. But, the idea of
ship—the first idea—was not so formed. Nor was the idea of ship
improved, Vico recalls, by a P. Perot, who, in Vico's time, built a
boat according to geometric principles. The boat sank immediately
upon launching.[27]

Familiar things can be analyzed by reason. Reason is at home in
the familiar. Reason, however, does not establish the familiar,
which it may recognize and organize. Reason, as Kant knew,
together with the understanding, aids in structuring the familiar,
although he does not state it in this way. That is, the concept—the
intelligible universal—provides an a priori level for experience.
But, reason and the understanding do not provide an original
grounding, Vico believes. The imagination provides that original
and originating ground that produces the familiar out of the unfamiliar, out of an experience of ultimate mystery.

In the *New Science* Vico creates a noble fiction—a humanistic fiction—a tale of the founding of society and culture. Vico's story begins with the recognition that the first humanity was gigantic and illiterate. Nonetheless, this race of barely human beasts managed to form societies and language. Vico is clear that the social contract theory, a fiction produced by such Cartesian mentalities as Grotius, Seldon, and Pufendorf, is wrongheaded, a pure fiction of science created simply out of abstract possibility.[28] Vico finds it unlikely that a humanity unable to speak and to write—able to form only animal sounds—would be able to draw up legal contracts.

Vico believes that those who constitute the first humanity, barely able to speak, wander in the forests where they are at the mercy of natural forces, which of course they do not understand. They are unable to form groups inasmuch as they lack language and a sense of place. Their sense of place is formed, Vico has it, in a linguistic act, when the noble poets among the first men name the thundering sky that so frightens all the rest. During great storms, humanity simply disperses. But, in one version of this originating poetic act, Vico writes:

> Thereupon a few giants, who must have been the most robust, and who were dispersed through the forests on the mountain heights where the strongest beasts have their dens, were frightened and astonished by the great effect whose cause they did not know, and raised their eyes and became aware of the sky. And because in such a case the nature of the human mind leads it to attribute its own nature to the effect, and because in that state their nature was that of men all robust bodily strength, who expressed their very violent passions by shouting and grumbling, they pictured the sky to themselves as a great animated body, which is that aspect they called Jove, the first god of the so-called greater gentes, who meant to tell them something by the hiss of his bolts and the clap of this thunder. And thus they began to exercise that natural curiosity which is the daughter of ignorance and the mother of knowledge, and which, opening the mind of man gives birth to wonder.[29]

The role of ignorance must be noted in Vico's epistemology. Vico writes of a natural curiosity that is produced from ignorance and yet is the mother of knowledge. The noble poets among the first humans name the sky in the presence of what they do not understand. Motivated by fear and powerful emotions, nonetheless they stand and name. In Vico's account of this first imaginative act, naming takes place in fear, spiritual agitation, and in the presence of a mystery before which one trembles. This name, Jove, is the first imaginative universal. From this account, Vico derives a principle in relation to poetic ignorance, which he understands as *"l'indiffinita natura della mente"*—an indefinite nature of mind. Vico states: "Because of the indefinite nature of the human mind, wherever it is lost in ignorance, man makes himself the measure of all things."[30] In another place Vico writes:

> So that, as rational metaphysics teaches that man becomes all things by understanding them (*homo intelligendo fit omnia*), this imaginative metaphysics shows that man becomes all things by *not* understanding them (*homo non intelligendo fit omnia*); and perhaps the latter proposition is the truer than the former, for when man understands he extends his mind and takes in the things, but when he does not understand he makes the things out of himself and becomes them by transforming himself into them.[31]

Vico acknowledges two directions of spirit. Reason moves outward toward the object in order to understand it. In its movement it creates intelligible universals out of what it understands and out of what is before it. The object is already familiar to it. Imagination, on the other side, takes in what it does not understand and fashions the imaginative universal out of itself. Knowing becomes an empathic exchange. The object for imagination is an alter-body of itself. Spirit only seems to move outward. The imaginative universal is embodied imagination. The first direction of spirit—reason— produces science; the second direction provides the humanities. The humanities provide true knowledge, Vico concludes, presuming the principle that the mind can best know what it has made.[32] Humanistic knowledge involves the mind knowing what it has made, and what it first makes are language and culture. The sciences, at best, can obtain certain knowledge but not the knowledge, in their attempts to know what God has made.

To accomplish humanistic making—true language and culture— Vico understands that human gestures had to learn to speak. The first humans, Vico claims, are barbarous because they think with their bodies—thus their giantism—and communicate in mute gestures. At first, humanity animates material substances with spirits; they project their own crude mentalities upon the body of the world.[33] Making the linguistic connection between *mythos* and *mutus*, Vico posits a prelinguistic and prereflective epistemological ground whereupon no bifurcation exists between subject and object, mind and world.[34] Vico posits this ground from his reading of Greek and Roman literature. For example, in the exchange between Idanthyrsus, king of the Scythians, and Darius the Great of

Persia, Vico finds his theory of mute gestures confirmed. Darius challenges Idanthyrsus to war and the Scythian king replies by sending Darius five objects—a frog, a mouse, a bird, a ploughshare, and a bow. Vico concludes that Idanthyrsus, with this language that is even more primitive than hieroglyphics, was claiming to be born of the earth, to make his home there, to have the auspices on this side, to cultivate that land, and to defend that land.[35] In the language of the mute sign, Darius had better leave or be killed.

Vico's own understanding is fantastic and is an example of what Donald Phillip Verene, in *Vico's Science of Imagination*, calls "recollective *fantasia*."[36] Vico's fantastic recreations are imaginative accounts of classical literature, which would not have satisfied Poe's yearning for verisimilitude. For Vico, the *verum factum* principle applies—the true is the made.[37] In Herodotus's account of the exchange between Darius and Idanthyrsus, Idanthyrsus responds with the following: "I will send you no gifts of earth and water, but others more suitable; and your claim to be my master is easily answered— be damned to you!"[38] After plotting a suitable strategy for dispatching the Persian invaders, Idanthyrsus sends Darius a bird, a mouse, a frog, and five arrows. Gobryas, one of Darius's counselors, interprets these signs: "Unless you Persians turn into birds and fly up in the air, or into mice and burrow under ground, or into frogs and jump into the lake, you will never get home again, but stay here in this country, only to be shot by Scythian arrows."[39]

Vico has the facts wrong but the story right, giving us its truth in an account that explains the totality of the experience. Vico has offered us Jove's vision: he constructs a total understanding out of isolated and disparate facts. He is like the noble poets among the first humanity who were able to imagine and then provide a shape for particular sensual experience. The noble poets among the first humans take an event consisting of sights, sounds, feelings, and ideas and give them a linguistic form. In this form the meaning transcends the experience. Jove is all that is and all that can be. The theories of the mute gesture and the imaginative universal explain what those people were able to form but not understand. A notion of universality precedes any abstract expression of it. The concrete particular in the here and now must come alive with a significance that transcends its simple appearance, its visual presentation. The thundering sky becomes a god for the first humanity. With this image—the first metaphor—they are able to form notions of other gods and other civil institutions. The mute gesture is the first step that leads to the metaphor, a totality that is a concrete universal, a synthesis of words and images.

Imagination, Vico contends, is compounded memory that shapes and alters experience. Memory does not simply recall. Vico writes: "Memory . . . has three aspects: memory [*memoria*] when it remembers things, imagination [*fantasia*] when it alters or imitates them. and invention [*ingenio*] when it gives them a new turn or puts them into proper arrangement and relationship."[40] Thus, the hiss and the flash of the lightning, the clap the thunder, the fear of the first humans, the insight of the poets, the linguistic act, and the immensity of the sky come together. The appearance of other gods is attendant upon the appearance of meaning itself, which is accomplished by the notion and image of Jove. In one other version of the story, Vico allows that Jove would most likely not have been the first word—more probably some expletive like "pa"—but for Vico these facts pale before their significance.[41]

In *memoria* the past is brought into the present. That which is not appears in relation to what is. The sensual moments that flow in time are stopped. Next, *fantasia* understands these moments in a linguistic form in which the radically new—the Wholly Other— appears. Finally, *ingenio* enables the poets and the people to rearrange their experience. Now all things are full of gods, and the people set out to name them. A *sensus communis* is established.

The establishment of community meaning requires an imagination that does what reason cannot do. Vico states: "It is impossible that bodies should be mind, yet it was believed that the thundering sky was Jove."[42] Reason judges in terms of consistency. Bodies cannot be minds. This is Descartes's judgment. If mind is unextended and thinks, then it cannot be body—that is, it cannot be extended. Nonetheless, in some sense body is mind, a reality that greatly troubled Descartes, who attempted to locate their interaction in the middle-landscape of the pineal gland. The imagination is not troubled by contradiction; the imagination in fact requires fundamental opposition in its working. What is becomes what is not. The familiar is made from the unfamiliar, which necessitates action and response. The *sensus communis* of the first societies is formed as the thundering sky is imaged forth as a great shaking and flashing body making pronouncements that become the basis for civil life. This first sensuous grasp of unity, of an all-together, provides the basis for conceptual and cognitive unities that are achieved through powers of abstraction. Again the two directions of spirit are affirmed in social, historical, and intellectual life.

For Vico social life requires a notion of Providence, a sense of meaningful ritual, and recognition of mortality—the fundamentals for a *sensus communis*. A culture becomes decadent, however, as the powers of *fantasia* wane. Vico says: "Men first feel necessity, then look for utility, next attend to comfort, still later amuse themselves with pleasure, thence grow dissolute in luxury, and finally grow mad and waste their substance."[43] Vico, in his own time, witnesses what he believes to be a corruption of the notion of Providence. He sees the family in disarray; he finds a general neglect of the past. He discovers a new barbarism far worse than that of the first people wandering in the feral forests. Reason provides the appearance of understanding but not its source. The new barbarism is a barbarism of reason, which Vico lays at the feet of Descartes. A failure of culture is a failure of the imagination. The arts become a barometer. When, for example, creation becomes a matter of technique, creation is no longer understood. Vico reminds us that technique, know-how, and inventiveness are the results of *fantasia*, which pretechnological people imaged forth with the imaginative universal of thrice-great Hermes. Invention does not, of necessity, lead to inventiveness. Vico states: "In every [other] pursuit men without natural aptitude succeed by obstinate study of technique, but he who is not a poet of nature can never become one by art."[44]

We learn from Vico that each age needs a sense of its own beginnings as well as a premonition of its own ends. A sense of Providence leads to the notions of death and goal. That is, each nation must understand itself in terms of an ideal eternal history, in which birth leads to decay and death. Eternal progress is an infernal fiction. A culture's inventiveness, its *sensus communis*, is at stake. What is and what is not are absolute oppositions that are never resolved; the activity of imagination depends upon it. A culture's metaphorical constructions are responses to the recognition of decay and death. The generating sense of absolute mystery is likewise never resolved.

Poe, as I have shown, is a man of his time, and his time is greatly interested in the imagination, perhaps sensing its demise. Poe is led in two directions. He tries to correct the allegorical abuses of writers like Hawthorne; he urges that poetry should respond to the exigencies of everyday life; he espouses the virtues of diligent work and methodology. Nonetheless, he writes, in the "Poetic Principle," that verse should never stray too far from the rhythm of the ear, that likely the *Iliad* was not properly conceived as a long epic poem but that it probably contained lyric dimensions lost to modern readers. Poe, in this essay, recalls the oral tradition of ancient poetry easily forgotten in the march of literacy. Science obscures, Poe will acknowledge, although in the end he succumbs to science's irresistible appeal. Of the imagination he writes:

The pure Imagination chooses, *from either beauty or deformity*, only the most combinable things hitherto uncombined;—the compound as a general rule, partaking (in character) of sublimity or beauty, in the ratio of the respective sublimity or beauty of the things combined—which are themselves still to be considered as atomic—that is to say, as previous combinations. But, as often analogously happens in physical chemistry, so not unfrequently does it occur in this chemistry of the intellect, that the admixture of two elements will result in a something that shall have nothing of the quality of one of them—or even nothing of the qualities of either. The range of Imagination is therefore, unlimited.[45]

In this nightmare of prose, scientific understanding wars with poetic sensibility. Poe somehow grasps that the imagination combines, that what is combined is originally in an atomic form, and that the combination appears as something new. He is aware that the imagination works in relation to a fundamental opposition—for example, beauty and deformity. Poe understands this combining in rather mechanical terms. Ratio and reason are active agents in his intellectual chemistry. Predictably, he regards the imagination as unlimited.

For Vico, the imagination functions only in relation to limits—absolute limits in which the ultimately meaningful is posited. The meaningful rests in the joining of contradictory elements. Reason works in relation to absolute limits that are presupposed. The law of contradiction assumes the relationship of identity and of the excluded middle. The notion of Jove, of the ultimately meaningful, makes all deductive logic possible. Analysis is endless but the ends of analysis are not achievable. Those ends, for Vico, are framed by the imagination. The meaningful is a cornucopia of meaning: in the absence of the meaningful, meaning cannot be obtained. Reason is unlimited because it can never provide its own justification. Absolute identity is only possible as a goal; the law of noncontradiction provides the motive for method but never its completion, as Hegel, that genius of reason, knew. The absolute, Hegel realizes at the end of *The Phenomenology*, is a calvary, a place of death, a testament to the failure of consciousness at becoming one with its object.[46] *The Phenomenology* pays homage to that failure, to the attempts of consciousness to achieve the absolute. But, where does the quest begin? For Vico, consciousness begins when the thundering sky speaks to the noble poets among the first people, who in their turn listen. The naming of the gods follows with the concomitant delineation of civil institutions, laws, and language.

Vico believes that each nation runs a course, ultimately ending in ruin. When a people becomes barbarous—losing their *sensus communis*—reason will not revive cultural social life. Such a barbarous state will either disappear, be conquered, or return to a new religious life. Vico writes: "Wherever a people has grown savage in arms so that human laws have no longer any place among it, the only powerful means of reducing it is religion."[47] Reason may masquerade as the imagination, as it likely does in Poe's case. As we recall, for Poe, beauty is embodied in the death of a beautiful woman in a poem 108 lines long, regularly and precisely interrupted with refrain "Nevermore." Beauty becomes whatever Poe thinks it is. And think, he does. He thinks in terms of clichés, in terms of the reader coming home after a hard day's labor, too busy and tired to wade through another Homeric epic, and in terms of scene shifting and tackle hoisting. A poem produced with the proper methods cannot fail to have a desired effect. The beautiful, in Poe's hands, becomes a cliché, a particular object, constructed by the machinations of reason. It is significant that Poe turns to mystery writing, to journalism, and to what some call science fiction. Religion is never far behind, in Vico's schema. When a people loses its sense of the beyond, which establishes culture, when ingenuity becomes ingenuous, then a form of religious barbarism—the worship of particularity—will return. The need for the sacred never subsides but returns in a new religion, which in the twentieth century is science and technology.

3

The images of the gods or a notion of a transcendental God no longer sustains our age. Rather, the conceptual verities of the natural sciences and the tautologies of mathematics provide the bases for our social life, or so it would seem. In his remarkable essay, "Flying Saucers: A Modern Myth of Things Seen in the Sky," published in 1958, C. G. Jung suggests that the gods have reappeared but that their form has been significantly and racially altered.[48] In the past humanity looked to the sky for reassurance and for warnings: accounts of circles and fireballs in the sky were, in primitive cultures, interpreted as signs from the gods. These "mandala-like" figures, symbols of unity and harmony, according to Jung's thesis, often appeared in times of cultural or personal crisis. Jung noticed that such sightings were on the rise in his own time, and he connected these with the social uncertainties over the cold war with Russia, over the distinct possibilities of nuclear war, and over the myopia resulting from a technological frame of mind. Jung writes:

It is characteristic of our time that the archetype, in contrast to previous manifestations, should now take the form of an object, a technological construction, in order to avoid the odiousness of mythological personification. Anything that looks technological goes down without difficulty with modern man.[49]

The explanation of the extraordinary is reduced to the technologically ordinary, albeit the ordinary from another world. Jung writes:

> In this matter there is little choice today: anything out of the ordinary can only be pathological, for that abstraction, the "statistical average," counts as the ultimate truth, and not reality. All feeling for value is repressed in the interests of a narrow intellect and biased reason.[50]

To determine whether a flying saucer sighting is a fact, witnesses and evidence are sought. In evaluting testimony, it is crucial to decide whether the sighting is a trick of the imagination. Jung argues, however, that the technological mentality that despises the imagination is likely to be avenged by it; the very people who serve as expert witnesses for saucer sightings, because they "were never distinguished for . . . [their] lively imagination or credulousness," are the very people who are most susceptible to such visions.[51]

Whether these phenomena are visitors from the stars is of little importance for Jung's case. The facts of the matter pale before their perception. That these phenomena are seen as interplanetary visitors is the critical issue. I do not necessarily follow Jung in taking these sightings as projections from a tortured unconscious, although there is certain music to his theory. I do, however, take them as evidence of a cultural need for mystery and a failure of imagination.

No clearer example of this failure can be found than the best seller by Erich Von Daniken, *The Chariots of the Gods? Unsolved Mysteries of the Past*.[52] This extraordinary book sold over three million copies and influenced an untold amount of television time. Von Daniken's thesis is simple: the mysteries of the past can be solved by appealing to technology. Consider the following argument:

> Where did the narrators of *The Thousand and One Nights* get their staggering wealth of ideas? How did anyone come to describe a lamp from which a magician spoke when the owner wished?
>
> What daring imagination invented the "Open, Sesame!" incident in the tale of Ali Baba and the forty thieves?
>
> Of course, such ideas no longer astonish us today, for the television set shows us talking pictures at the turn of a switch. And as the doors of most large department stores open by photocells, even the "Open, Sesame!" incident no longer conceals any mystery.[53]

Mystery is thus resolved as the flick of a technological switch. In this argument, and throughout the book, the assumption is made that the nature of the human is fundamentally technological. If the first people, the implied argument goes, were truly human and able to do marvelous things, they must have had access to advanced technology; therefore, they were visited by a technologically superior race that got them started.

Technology and the miraculous are joined in Von Daniken's mind and, likely, in the sensibilities of the great mass of readers. Von Daniken's book cannot be dismissed, ridiculed merely as popular sensationalism that brought little serious and learned attention. The link of technology with the sacred and the equation of the imagination with a methodical, problem-solving mentality is greatly significant.

Clearly technology has co-opted the sense of the possible and the range of absolute limits. Physics professor Freeman Dyson, at Princeton's Institute for Advanced Studies, in his 1985 Gifford lectures, writes: "Technology is a gift of God. After the gift of life it is perhaps the greatest of God's gifts. It is the mother of civilizations, of arts and of the sciences."[54] Like Von Daniken, Dyson sees technology as a creative force, even more fertile than science, as basic as God-given life itself. Dyson's view of life moves well beyond common sense, or rather it extends the new common sense, a denial of sense. Dyson understands life in true technical fashion, as organization. He writes: "If this assumption is true that life is organization rather than substance, then it makes sense to imagine life detached from flesh and blood and embodied in networks of superconducting circuitry or in intersteller dust clouds."[55]

We are asked to move life to the level of conceptual construction, which Dyson takes to be more basic than life of flesh and blood. Of course, reason may dissect any whole to constituent parts. The parts, however, do not exist as parts, as elements, but as atomic elements, that individually do not exist. A group may be described in terms of rules and order, but the description does not have the same reality as the group to which it applies. The notion of organization that reason conjures, though, may be taken as real, committing what Whitehead called the fallacy of misplaced concreteness and perhaps what Vico understood as the barbarism of reason. For Dyson, the substance of life becomes its organization, a view that moves the ivory tower well below street level, a view that is employed every day by corporate management and by kitchens of fast-food restaurants. Possibility and actuality are defined conceptually, if we press this assumption. What is technically possible *is* possible and, therefore, actual, given the proper problem-solving imagination. The intelligible universal takes on a life of its own.

The problems that technology creates can be handled by genetic engineering, Dyson reasons: "The scavenger turtle with diamond-tipped teeth, a creature programmed to deal in a similar fashion with human refuse and derelict automobiles," can be created and dispatched to help clean up the environment.[56] Logical possibility moves to reality. Logical possibility devoid of content is unimaginable. Logic takes over for the imagination. Science fact is indistinguishable from science fiction.

It is no easy matter to define science fiction, although critic Leslie Fiedler reminds us that publisher Hugo Gernsback gave the name to the genre, first calling it "scientifiction," and then later calling it "science fiction."[57] Gernsback founded the popular magazine *Amazing Stories* in 1926. He proposed a definition of the genre as well, which Fiedler states as: "A charming romance intermingled with scientific fact and prophetic vision."[58]

Gernsback was the king of what came to be called pulp science fiction that attracted a large popular audience as well as a dearth of popular criticism. *Fanzines* appeared, which invited professional contributions but was mainly the work of amateurs. Fiedler notes some of the variant definitions, which these enthusiasts devised:

> They range from statements like "the genre of literature that shows the limits and possibilities of man and goes beyond them"; and "the genre-... [which] explores the mental bush country beyond the tilled lands down by the old mainstream" to remarks like "Science fiction deals with everything that does not exist" and "Science fiction is what we mean when we point to it."[59]

For this popular audience, science fiction is whatever one thinks it is. No clear standards exist. With a certain sadness, Fiedler notes the intellectually respectable new-wave critics—Darko Suvin, Eric S. Rabkin, and Robert Scholes—and their attempts to tighten conceptual belts:

> My heart sinks at the phrase "serious readers," and when Rabkin and Scholes go on to say of the same novel that "this is not fiction for adults," I lose all hope. It is, after all, the availability to children and the childlike in us all, along with the challenge to the defunct notion of the serious reader which characterizes not just Van Vogt but all sf [science fiction] at its most authentic. . . . Any bright high school sophomore can identify all the things that are *wrong* about Van Vogt, whose clumsiness is equaled only by his stupidity. But the challenge to criticism which pretends to do justice to science fiction is to say what is *right* about him: to identify his mythopoeic power, his ability to evoke primordial images, his gift for redeeming the marvelous in a world in which technology has preempted the province of magic and God is dead.[60]

Fiedler understands that beneath the seemingly endless diversity of science fiction writing is the demand for myth and mystery. The business of science fiction, and perhaps all prose, is not to satisfy the demands of consistency but to plumb the depths of the inconsistent where the lands of myth lie. Fiedler notes the work of Brian Aldiss, *Billion Year Spree*, which tries to be less elitist, less exclusive. The task is difficult, Fiedler admits, cognizant that even the best of science fiction is not likely to be on the same level of modern writers like Joyce, Proust, Mann, or Kafka. Science fiction is a kind of "sub- or para-literature."[61] Fiedler writes:

> Indeed there is a clue to this [in the mystery of genre identification] in the metaphor Aldiss uses in praising *Star Maker*, which he describes as "really the one great grey holy book of science fiction."
>
> "Holy" is the operative word, suggesting that we do not read *Star Maker*—if we are responding to it properly—as we read, or, God forbid, analyze or interpret James Joyce's *Ulysses*. We experience it "scripturally," as perhaps unintended but quite valid "Scripture," that is, as the mythological statement, accepted on faith, of a grid of perception through which we see or aspire to see the world and ourselves. Much popular literature . . . is read thus; and therefore eventuates not in ethical enlightenment or a heightened appreciation of virtuosic form, but in transport, ecstasy, a radical transformation of sensibility—thus permitting us to transcend momentarily the normal limits of consciousness and escape the ennui of routine awareness, the restrictive boundaries of the unitary ego.[62]

Although Fiedler does not pursue it, there is a twofold theory of transcendence at work in this discussion. The science fiction text, in this case Olaf Stapledon's *Star Maker*, allows for momentary transcendence of "routine awareness." This text is a kind of scripture, a grid, through which we see the world. I take this to mean that there is an identification with the text, such as does not occur with the works of Joyce, which often distance the reader. And then there is the sense of transcendence that provides ethical and aesthetic enlightenment and a transformation of sensibility. As I understand it, the first sense of transcendence is very much tied to the mundane, which, as I am arguing, is the technologically familiar. Fiedler acknowledges the need to move beyond the technological, and perhaps science fiction is such an attempt, although, to use Poe's terms, I fear that science fiction is little more than scene shifting and tackle hoisting, however pleasurable those things might be. Transcendental success, I believe, would require the second sense of transcendence, would require what Vico would understand as *fantasia* rather than *ragione*, reason, in disguise.

It is significant that science fiction aficionados single out British philosopher Olaf Stapledon's work as sacred. In writing of his vision of the future in *Last and First Men*, in 1930 Stapledon writes:

> Our imagination must be strictly disciplined. We must endeavour not to go beyond the bounds of possibility set by the particular state of culture within which we live. The merely fantastic has only minor power. . . . The activity that we are undertaking is not science, but art; and the effect that it should have on the reader is the effect that art should have.
> Yet our aim is not merely to create aesthetically admirable fiction. We must achieve neither mere history, nor mere fiction, but myth. A true myth is one which, within the universe of a certain culture (living or dead), expresses richly, and often perhaps tragically, the highest admirations possible within that culture. . . . This book can no more claim to be true myth than true prophecy. But it is an essay in myth creation.[63]

Stapledon affirms the importance of myth, although he allows that his book is only an essay in myth creation. He is employing a "disciplined imagination." He says that his concerns are aesthetic, that he considers effects—embodying consciously or not Poe's dicta. And, he is careful, he says, not to stray too far from the verisimilitudes of his age.

Facing in part those truths, Stapledon writes:

> We all desire the future to turn out more happily than I have figured it. In particular we desire our present civilization to advance steadily toward some kind of Utopia. The thought that it may decay and collapse, and that all its spiritual treasure may be lost irrevocably, is repugnant to us. Yet this must be faced as at least a possibility. And this kind of tragedy of a race, must, I think, be admitted in any adequate myth.[64]

Any attempt to classify science fiction, to understand what it is and what it is not, must come to terms with the notion of utopia. In the above remarks Stapledon is simply saying that it has become commonplace to think of civilization as advancing, along with technological development, toward the better life.

Progress and civilization go hand in hand. Stapledon, writing in the 1930s after World War I, knows better. The *Last and First Men* is a dystopian or antiutopian novel, a novel of tragedy, although in the previously quoted preface, Stapledon cannot hold back hope. His fanciful history is a history of despair and tragedy, where rationality is misapplied. In some measure he hopes his book to awaken humanity to the imminence of tragedy: "May this not happen! May the League of Nations, or some more strictly cosmopolitan authority win through before it is too late!"[65] Through the despair, reason still shines. Stapledon never seriously questions the efficacy of reason, although he tries to explore its limits, for example, in imagining a race—the Third Men—which is telepathic and which does not have to rely on words but can convey thought itself. In the main, the refrain is that thought is misapplied.

Robert M. Philmus, in "Science Fiction: From its Beginning to 1870," finds utopia and dystopia as main topics.[66] These *topoi* help to locate the vast region that is potentially science fiction. Thomas More's *Utopia* of 1516 is thus a landmark work, coining the word utopia, meaning "no place" and punning on "the good place." Thus, from Philmus's point of view, Tommaso Campanella's *City of the Sun* (1623) and Francis Bacon's *The New Atlantis* (1627) would be included. Also considered are Cyrano de Bergerac's *The Comical History of the States and Empires of the Worlds of the Moon and Sun* (1651) and Denis Diderot's *Supplement to Bougainville's "Voyage"* (1772). In all of these cases the problem of place is explored. Particularly of interest in Philmus's bibliography is the role reason plays in establishing place, at least on my reading of his work. Reason may be attacked and ridiculed but it is always taken seriously, especially in its relation to technological advancement. Thus, it should be no surprise that utopian ideas are expressed in the early modern era, during the rise of national, scientific, and production interests.

Philmus notes Plato's *Republic* and his mention of Atlantis in the *Timaeus* as utopian interest, although he does not include these works in his bibliography, and I think for good reason. Plato believes the Republic, as an actual state, could never see the light of day. Reason shows that the existing state is a corrupt state (bks. 8–9). Plato is a philosopher of reason but no friend of the state, no believer in the practical efficacy of reason.

Utopian interest blossoms, however, in the rise of the nation state and in the application of technology to national power. Science fiction is witness to this rise, as is the development of popular literature.

4

Literacy and the power of the state go hand in hand, contemporary French critic Jacques Ellul argues. Technology, a mentality seeking the efficient as an absolute methodology in all areas of social life, requires a literate citizenry, able to be propagandized. By propaganda, Ellul does not mean simply the inflammatory rhetoric of an authoritarian state. In his landmark studies, *The Political Illusion* and *Propaganda*, Ellul argues that the business of the state, dependent upon technical development as the locus of power, is the creation of a mass sensibility, which may be translated as a willingness to reason.[67] That is, the citizen must come to see his or her own world as fictional, must come to see ultimate reality in terms of state interest and in terms of scientific and technological development. In short, for the modern state to prosper, the modern citizen must come to live in a world of mass media, in a world of the image and the concept.

The citizen may come to trust in the news that he or she reads or the citizen may come to disbelieve anything that is read or heard. The results are somewhat the same. The individual's immediate and bodily world is called into question. The individual is alienated and alone. To be alone is to need comfort. Ellul writes:

> Utopia is the "Negro spiritual" of modern intellectuals of the West. It is a consolation in the face of slavery, an escape from something one is unable to prevent, a spiritual dimension, a separation of a free intellect from an enslaved body, a reinforcement of faith.[68]

The notion of utopia is what we whistle while we work. It is the belief in progress, in the capacity of reason and technology to solve the problems that reason and technology create. But the idea of utopia may also be what diverts us from having to consider the concrete reality in which we barely live.

To understand technology as having to do merely with machines is not to understand it at all. Machines can be unplugged and to some degree controlled. The machine, Ellul admits, is a model, but one that in the technological society is superseded. All that technique touches becomes machinelike, Ellul contends; with the application of rationality and judgment, the technical phenomenon is the result.[69] The technical phenomenon is technical reason embodied. This embodiment occurs as a technical operation, typically a bodily activity like chopping down a tree, working a keyboard, or walking to the store on an errand, is surveyed with an eye to performing it the "one best way," adopting a mathematics-like methodology. A bulldozer, some computer software, or a delivery service—all technical phenomena—might be the results. Technical consciousness or technical intention appears when all areas of social life are thus scrutinized and modified. Everything must be done the one best way. At this stage, technical consciousness is out of control, unable to recognize itself as consciousness, no longer in touch with the object of its own activity.

Technique becomes a self-augmenting system, automatically chosen, universally adopted, and proceeding with absolute causal connection. Any choice that is made is made technically, automatically. Technique becomes monistic—its own world—when that which can be done will be done. All choices are linked. And technique becomes the sacred, that which is ultimately deserving of respect. The technical phenomenon is originally the result of a rational distance on the world, in which consciousness stands before an object and goal to be accomplished. As the one best way is sought, adopting methods of rational efficiency, all other approaches and methodologies are excluded.[70] H. L. Menken's advice that anything worth doing is worth doing badly could not be heard.

The technical phenomenon appears when the rational becomes the real—embodied or not as a machine—and when methodology triumphs over spontaneity. Poe's poetic dicta may function as a technical phenomenon. The technical phenomena clarify Vico's notion of the intelligible universal, an example of its capacity to have a life of its own. The technical phenomenon as an intelligible universal transforms the vision and the understanding of the culture in which it is embodied, producing a form of barbarism Vico could only imagine. This new barbarism is, of course, a theme of much early science fiction.

Samuel Butler, in *Erewhon* (1872), imagines a society that has banned the machine but not its logic. In the famous chapter "The Book of the Machines," the arguments leading to the banning of the machine are presented. Adopting a version of Darwinian thinking, an Erewhonian prophet writes that the line between the conscious and the unconscious, between the organic and the inorganic, is arbitrarily drawn. Every organic thing can be reduced to inorganic things. Every whole can be divided into parts. Therefore every whole and every part may somehow be logically related.

> Either . . . a great deal of action that has been called purely mechanical and unconscious must be admitted to contain more elements of consciousness than has been allowed hitherto (and in this case germs of consciousness will be found in many actions of the higher machines)—or (assuming the theory of evolution but at the same time denying the consciousness of vegetable and crystalline action) the race of man has descended from things which had no consciousness at all. In this case there is no *a priori* improbability in the descent of consciousness (and more than conscious) machines from those which now exist, except that which is suggested by the apparent absence of anything like a reproductive system in the mechanical kingdom. This absence however is only apparent, as I shall presently show.[71]

Machines do reproduce in that they make humanity dependent upon them, or perhaps they introduce humanity to this need for dependence. In any case, machines serve so that they may rule, Butler understands, by reducing humanity's vision to the material level, reducing consciousness to repeatable axioms, and chaining reason to the laws of contradiction.[72] The Book continues:

> Man's very soul is due to the machines; it is a machine-made thing: he thinks as he thinks, and feels as he feels, through the work that machines have wrought upon him, and their existence is quite as much a *sine qua non* for his, as his for theirs. This fact precludes us from proposing the complete annihilation of machinery, but surely it indicates that we should destroy as many of them as we can possibly dispense with, lest they should tyrannize over us even more completely.[73]

This prophet recognizes the degree to which machines create a system. They enable us to produce more so that we may eat more. We eat more; we have more children; we work more: therefore, we need more machines. Technology does not save labor but rather creates labor on its own terms: "The machines being of themselves unable to struggle, have got man to do their struggling for them."[74]

The Erewhonian society bans machines only to fall prey to be the thinking that makes them possible. Thought becomes more sophisticated, more abstract, as the notion of machines as a form of life is entertained. Now, all things may be considered alive. Life is what thought can conceive. The rights of animals are maintained by a specific prophet, whose thought becomes law. Of this prophet, Butler reveals:

> All prophets are more or less fussy, and this old gentleman seems to have been one of the more fussy ones. Being maintained at the public expense, he had ample leisure, and not content with limiting his attention to the rights of animals, he wanted to reduce right and wrong to rules, to consider the foundations of duty and of good and evil, and otherwise to put all sorts of matters on a logical basis, which people whose time is money are content to accept on no basis at all.[75]

It is interesting that Butler describes Erewhonian social change in relation to prophetic insight, that he sees the machine to be an extension of a form of consciousness, and that he anticipates the spread of this consciousness systematically throughout the social order. On the one hand, change is shown to be a logical progression; on the other hand, change is rooted in a kind of divine insight, a penetration into the realm of mystery, requiring prophetic vision.

Ellul suggests that technology establishes a sacred order when it is realized that science and technology cannot provide a society with a sense of being and purpose, meaningful ritual and a sense of place. The need to transcend the rational appears. Ellul writes:

> Here is man in his new milieu, in our modern society, feeling alone all over again, and without a respondent. He directs at the empty sky a discourse without dialogue. He embraces things, which are never anything but objects. He is living anew the horror of silence and incompleteness.[76]

Surrounded by images, by facts, by silent machines and efficient methodologies, humanity looks again to the sky, to the word, to meaning that surrounds. Science fiction and mystery stories are discourses addressed to this silence, to this emptiness. Needed, however, is a literature or perhaps a preliterature of the imagination espousing a new sensibility, a new language, and a proper respect for mystery that is something more than the reification of the familiar.

Mystery does not invite mere obeisance and servitude, scene shifting and tackle hoisting. It requires a true effort, a labor of the negative from which whole cultures are the result. Humanists, the properly respectful but not overly servile minions with visions of the Wholly Other, must not abdicate their duties to science and technology. They must work narrations of the whole in metaphors that continue what the first humanity attempted in its expressions of Jove.

NOTES

A much abbreviated version of this essay appeared in *The Nebraska Humanist* 3 (Spring–Fall 1981): 36–45. The idea for the essay began during a 1981 NEH Summer Seminar at Michigan State University. Seven years later, the title and the general idea of the essay—and little else—remain the same.

1. *The Technological Society*, trans. John Wilkinson (New York: Alfred A. Knopf, 1964), p. 43.

2. *Discourse on Method*, in the *Philosophical Works of Descartes*, trans. and ed. Elizabeth Haldane and G. R. T. Ross (New York: Dover Press, 1955), 1: 87–93.

3. Giambattista Vico, *On the Study Methods of Our Time*, trans. Elio Gianturco (New York: Bobbs-Merrill, 1965).

4. Ibid., pp. 21–44.

5. "Edgar Allan Poe, His Life and Works" (1852), in *Affidavits of Genius: Edgar Allan Poe and the French Critics, 1847–1924*, ed. Jean Alexander (Port Washington, N.Y.: Kennikat Press, 1971), p. 115.

6. Edgar Allan Poe, "The Philosophy of Composition," in *Literary Criticism of Edgar Allan Poe*, ed. Robert L. Hough (Lincoln, Neb.: University of Nebraska Press, 1965), p. 23.

7. Ibid., p. 21.

8. Edgar Allan Poe, "The Murders in the Rue Morgue," in *The Complete Tales and Poems of Edgar Allan Poe*, introd. Hervey Allen (New York: The Modern Library, 1938), p. 143.

9. Ibid., p. 157.

10. Poe, "The Poetic Principle," in *Literary Criticism*, p. 39.

11. Ibid., p. 38.

12. Ibid.

13. Ibid.

14. Poe, "The Philosophy of Composition," in *Literary Criticism*, pp. 231–41.

15. Poe, "The Poetic Principle," in *Literary Criticism*, pp. 24–25.

16. Poe, "Tale-Writing: Nathaniel Hawthorne," in *Literary Criticism*, p. 147.

17. Edward H. Davidson, *Poe: A Critical Study* (Cambridge: Harvard University Press, 1957), pp. 223–53.

18. Daniel Hoffman, *Poe Poe Poe Poe Poe Poe* (New York: Doubleday, 1972), p. 87.

19. Remy De Gourmont, "Marginalia on Edgar Poe and Baudelaire," in *Affidavits of Genius*, p. 232.

20. Poe, "The Imp of the Perverse," in *The Complete Tales*, p. 282.

21. Immanuel Kant, *Critique of Judgment*, trans. J. H. Bernard (New York: Hafner Press, 1951), p. 100.

22. Ibid., p. 105.

23. Rudolph Otto, *The Idea of the Holy*, trans. John W. Harvey (Oxford: Oxford University Press, 1950), p. 42.

24. Ibid., p. 44.

25. For the most complete discussion of Vico's notion of *fantasia*, see Donald Phillip Verene, *Vico's Science of Imagination* (Ithaca: Cornell University Press, 1981), esp. chap. 3.

26. Giambattista Vico, *The New Science of Giambattista Vico*, trans. Thomas Goddard Bergin and Max Harold Fish (Ithaca: Cornell University Press, 1968), pars. 34, 209, 409, 460, 463, 501, 704, 934, 1033, 1040.

27. Vico, *On the Study Methods of Our Time*, p. 29.

28. Vico, *New Science*, pars. 310–29.

29. Ibid., par. 377.

30. Ibid., par. 120.

31. Ibid., par. 405.

32. I refer, here, to the *verum factum* principle established in Vico's *De antiquissima Italorum sapientia*, forthcoming from the Cornell University Press and translated by Lucia Palmer. See, however, Leon Pompa's partial translation in *Vico: Selected Writings* (Cambridge: Cambridge University Press, 1982), pp. 50–52.

33. Vico, *New Science*, par. 405.

34. Ibid., par. 401.

35. Ibid., pars. 48, 99, 435.

36. Ibid., par. 99.

37. Ibid., par. 405.

38. Herodotus, *The Histories*, trans. Aubrey Selincourt and rev. A. R. Burn (Baltimore: Penguin Books, 1954), p. 313.

39. Ibid., p. 314.

40. Vico, *New Science*, par. 819.

41. Ibid., par. 448.

42. Ibid., par. 383.
43. Ibid.
44. Ibid., par. 213.
45. Poe, "On Fancy and Imagination," in *Literary Criticism of Edgar Allan Poe*, p. 15.
46. G. W. F. Hegel, *Phenomenology of Spirit*, trans. by A. V. Miller with analysis and foreword by J. N. Findlay (Oxford: Clarendon Press, 1977), p. 493.
47. Vico, *New Science*, par. 177.
48. C. G. Jung, *Collected Works: Civilization in Transition*, vol. 10, trans. R. F. C. Hull (Princeton: Princeton University Press, 1964), pp. 309–433.
49. Ibid., p. 328.
50. Ibid., p. 338.
51. Ibid., p. 320.
52. Erich von Daniken, *The Chariots of the Gods? Unsolved Mysteries of the Past* (New York: Bantam Books, 1970).
53. Ibid., p. 65.
54. Freeman Dyson, *Infinite in All Directions: 1985 Gifford Lectures* (New York: Harper and Row, 1988), p. 270.
55. Ibid., p. 107.
56. Ibid., p. 155.
57. Leslie Fiedler, "The Criticism of Science Fiction," in *Coordinates: Placing Science Fiction and Fantasy*, ed. George E. Slusser, Eric S. Rabkin, and Robert Scholes (Carbondale and Edwardsville: Southern Illinois University Press, 1983), p. 2.
58. Ibid.
59. Ibid., p. 3.
60. Ibid., pp. 10–11.
61. Ibid., p. 12.
62. Ibid.
63. Olaf Stapledon, *Last and First Men and Star Maker* (New York: Dover Publication, 1968), p. 9.
64. Ibid., p. 10.
65. Ibid., pp. 10–11.
66. Robert M. Philmus, "Science Fiction: From its Beginning to 1870," in *Anatomy of Wonder: Science Fiction*, ed. Neil Barron (New York: R. R. Bowker Co., 1976), p. 11.
67. Jacques Ellul, *Propaganda: The Foundation of Men's Attitudes*, trans. Konrad Kellen and J. Lerner (New York: Alfred A. Knopf, 1967); *The Political Illusion*, tans. Konrad Kellen (New York: Alfred A. Knopf, 1967). Of particular interest in *The Political Illusion* is chapter 3, "Politics in the World of Images."
68. Jacques Ellul, *The New Demons*, trans. C. Edward Hopkin (New York: The Seabury Press, 1975), p. 117.
69. Ellul, *The Technological Society*, pp. 19–22.
70. See my "Jacques Ellul and the Logic of Technology," *Man and World: An International Philosophical Review* 10 (1977): 251–72.
71. Samuel Butler, *Erewhon* (New York: A Signet Classic, New American Library, 1960), p. 176.
72. Ibid., pp. 180–81.
73. Ibid., p. 180.
74. Ibid., p. 181.
75. Ibid., p. 200.
76. Ellul, *The New Demons*, p. 130.

Issue and Presentation: Technology and the Creation of Concepts

GAYLE L. ORMISTON

The investigations conducted by Monsieur G, the Parisian prefect, and his cohort in Edgar Allan Poe's "The Purloined Letter" supply an interesting analogy for understanding certain practices that fall under and dominate the heading of "science and technology studies." Monsieur G and his accomplice consider

> only their *own* ideas of ingenuity; and, in searching for anything hidden, advert to only the modes in which *they* would have hidden it. They are right in this much—that their own ingenuity is a faithful representative of *the mass*; but when the cunning of the individual felon is diverse in character from their own, the felon foils them, of course.[1]

For Monsieur G and his accomplice, then, there can be no variation in the principle(s) guiding their investigation. When an unusual circumstance or set of conditions arises, the old and accustomed modes of practice (and values) are exaggerated; the unusual or unexpected is rendered familiar through the application of specific principles that orient a particular method of inquiry. And yet, the principles remain untouched, idolized; the principles remain intact. However, without questioning the principles, neither the prefect nor those inquiring of "science and technology studies" can comprehend the radically different and differentiating turns, appearances, and myths encountered in the course of their investigations.

Under these conditions, the prefect/philosopher/STS analyst is guided by certain presumptions: (1) that some*thing* is concealed, hidden, or lost, where that which is lost or, at least, absent will be recognized or exposed; and (2) that identifying that which is absent will involve retrieving it along with a certain set of values or practices. To the extent that the prefect/philosopher/STS analyst does not question the principles and values that inform a particular investigation, he or she is and will remain deluded by the very application of those principles.

Today, even though we encounter the effects of technology in every attempt to comprehend "it"—that is, in every attempt to say with some clarity and certainty why the so-called question of technology bears any significance whatsoever—technology, like so many other concepts and practices, is pursued in a manner that resembles the method of Monsieur G. Like the "purloined letter," technology is "there"; but it cannot be represented as such, it cannot be located in any one place. Thus, technology appears, according to the most diverse accounts, to be a mystifying issue. Ubiquitous and mediate, yet never "present" as such, it remains suspended, withdrawn into the structures and practices of language, thought, and action. Only the application of certain principles, supposedly, will make it comprehensible. And yet, technology eludes apprehension. But why? Rather than attribute the status of autonomy to "technology," I would like to extend the analogy between investigative procedures by claiming that like "the cunning of the individual felon" whose character differs from that of Monsieur G. and his accomplice, and thus "foils them," "technology" does not match or correspond to the conceptual categories and oppositions, nor does it accommodate the principles, that direct the cunning of science and technology studies. In other words, there is a difference—always—between a word (a name) and the object(s) it signifies. There is a difference—always—between a concept and the practices it nominates. The difference, then, between the categories and principles that inform any method of inquiry (concerned with technology) and those "things" and "practices" that go by the name "technology" is the *issue of technology*.

2

In the likeness of the subject it treats, *the issue of technology*, what I present here is expectant, and prospective. Like any assay— or what Nietzsche calls a *Versuch*,[2] an attempt to "experience" a thought—this discourse looks ahead, beyond the immediate sensible presence of what goes by the name technology, to plot another way of talking about "technological presence." The discourse presented here, then, attempts to frame the issue of technology as the "dissemination" of techniques. It attempts to understand the issue of technology in terms of the production of intermediary concepts and practices—a self-supplementing chain of names, concepts, and practices into which "technology" is incorporated already. As such, this discourse attempts to understand how technology conditions and makes possible the translation of its conceptual and practical effects.

However it is presented or deliberated, the issue of technology is never, nor has it ever been, merely one question or problem among others. The word technology, the assemblage of concepts and the equipment and skills associated with and that inform the use of that term, designates both a problem and a field of interrogation. Not only does the issue of technology pervade all aspects of daily life, it penetrates the boundaries of the most diverse and heterogeneous disciplines and disciplinary discourses. Moreover, however the topic is presented or deliberated, technology is presupposed in and legitimates all questioning—all forms of fabrication.[3]

To advance such a hypothesis runs the risk of asserting an interesting yet obvious and trivial truth. If one comprehends "technology" in its most common sense, as the application of ideas or routine procedures in a specific context for the purpose of problem solving, or accomplishing certain tasks at a preestablished level of competence or expertise, then to say "technology is presupposed in and legitimates all questioning" is to say nothing new. Even though its observation remains seemingly obvious, the assertion itself is a fundamental assumption that informs the perspectives of much of the literature that traces the history of technology, the most notable of which is the text by Singer, Holmyard, and Hall.[4] Technology underlies all human history; in concert with its own development technology has guided the development of the epochs and episodes registered in our historical and analytical narratives, and continues to do so. The history of humankind—groups, societies, cultures—is a history inscribed in and through the implementation of artifacts, instruments and machinery of some sort, toward the organized control and manipulation of the environment, "natural" or "man-made."[5]

3

Etymological and lexical deliberations trace histories as well and contribute in many cases to the orientation of a particular discourse or narrative. Providing an ensemble of historical and conceptual accounts regarding the use of a word or a concept, etymologies are interwoven with and thus supplement that genre of text that would account for technological development in terms of how it has been a benefit or hindrance to the "emergence of man."[6] To be sure, a great deal of what is presumed in the contemporary discourse and analysis of technology studies—broadly construed to include philosophies of technology, sociologies of technology, politics of technology, histories of technology, and so on—is conditioned by an already *fixed* identity of technology, its "standard usage." For any analysis or discourse to proceed in such as fashion is to do so on the basis of having accepted *already* a concept of technology that is marked off by certain boundaries. Moreover, such usage not only assumes that the boundaries can be circumscribed or known but that various "technologies" or "techniques" share certain characteristic features that traverse and can be retrieved across contextual boundaries. The word technology is used *as if* it extended over a field of well-defined "things" and "procedures." Given this set of conditions, one can ask questions apropos "technology" similar to those Wittgenstein poses with regards to "games" and "language": "How is the concept of technology bounded? What counts as (a) technology and what no longer does?"[7]

Like an appeal to the "history" of the tool, where the development of certain practices and instruments discloses discrete changes in the use of some "thing," an appeal to the etymological derivations of a term or a concept sometimes will indicate ways for comprehending subtle differences in use, deployment, and significance. But are these differences is use not incorporated already into the lexical narrative that traces the different usages in which a term or concept has been rendered? Do these differences not determine, *already*, any understanding of the concept or use of the word technology? Do the differences in use, traced by an etymological and lexical account, not create or generate the possibilities for understanding the concept today, for rethinking and rewriting according to a different set of contextual constraints? The point of posing these questions is to draw attention to two statements, both taken from *A History of Technology*, regarding what "technology should mean":

Etymologically 'technology' should mean the systematic treatment of any thing or subject. In English it is a modern (seventeenth-century) artificial formation invented to designate systematic discourse about the (useful) arts. Not until the nineteenth century did the term acquire a scientific content and come ultimately to be regarded as almost synonymous with 'applied science'.[8]

And:

Technology should mean the study of those activities, directed to the satisfaction of human needs, which produce alterations in the material world. In the present work the meaning of the term is extended to include the results of those activities.[9]

To announce *what* " 'technology' should mean," or *that* " 'technology' should mean" is to open an inquiry in response to a specific question. To be certain, the question that can be answered in this fashion is "What is technology?" The question anticipates not merely a response but a certain *kind* of response—one that fashions the kind of response that will satisfy its demand. However, like any question taking this form, it anticipates. It is a question whose possibility depends on the presentation of an "object," or the representation of some "thing" called content. Here the form "what is . . . ?" functions only if a fixed identity for technology is presumed.

Of course, the cursory appeal to the etymology of "technology" that appears in the introductory comments of *A History of Technology* cannot be taken as a final statement about how technology is viewed throughout that text. Indeed, it must be understood that the appeal itself is made in the spirit of offering a "preliminary response" to the question "What is technology?" As its editors note, "It would hardly be useful to consider a term the range of which it is the purpose of the work itself to indicate."[10]

Nevertheless, despite the question's preliminary status, despite the provisional character of the response, the invocation of an *etymology*—itself a process of tracing, describing, and creating a discourse on "technology"—demonstrates how technology prefigures and circumscribes the issue of (its) definition. This appeal to an etymology, also demonstrates how the determination that "technology *should mean*" anything at all is not derived from a historical rendering or tracing of its uses and root terms, which one might expect from such a deliberation. Instead, this determination emerges from the *use* made of the etymological and lexical accounts required to satisfy a specific question: "What is technology?" To provide an answer to this question, both the history of the term and the history of the "tool" that term represents *should be* comprehended in a specified way. Rendering such a determination requires a crossing-over of thought and discourse from one universe of discourse to another, from the descriptive to the prescriptive. It requires an antecedent *trans*lation, a creation of names and concepts that places "all thinking on the subject" in a certain position, and thus stages the topical and thematic orientation of a discourse on technology.

4

To announce that "technology is presupposed in and legitimates all questioning" is to force a shift in the discourse on technology. It is, in an initial and provisional fashion, an attempt to frame the issue. By drawing attention to the ways by which technology is determined, to technology as a medium—a complex network of interlaced paths—for constructing different strategies and ways for questioning the uses and ends toward which technology is deployed, and to the values that underlie that deployment, attention is shifted toward the *ubiquity* and *mediacy* of technology. Technology appears at once as an artifact of human existense *and* as the medium by which the habitat of human existence comes to be what it is. To say that "technology is presupposed in and legitimates all questioning" is to recognize the mutual supplementation of practice and product —the structuring processes that mediate the relation between artifice and artifact, the joint articulation of concept and instrument. It represents an attempt to rethink and think through the concept of technology, to comprehend how the concept of technology, whatever its determination, mediates and thus becomes the condition for one's comprehension of "self" and "others." The process thus makes possible the habitat in which this comprehension is fabricated. To announce that "technology is. . . ." (in whatever context, for whatever purposes) is to create an-other concept, to render an-other translation, which is, in effect, to inaugurate a continued rethinking and rewriting of the concept against a background of undefined and indeterminate specifications. It is only on this basis that the translation from the descriptive account of use, provided by etymological and lexical narratives, to the prescriptive account of how a word "should be" understood, for instance, is made possible and legitimated.

To announce that "technology is presupposed in and legitimates all questioning" places the concept, and our so-called standard use of the term, at stake: the announcement throws into question "understanding," "consensus," "rational discourse"—the boundaries and rules within which the discourse or the analysis of technology makes sense—and necessarily so. For once the boundaries of any discourse have been stabilized, fixed by an antecedent translation of the question, or fixed by an understanding of a concept as a *whole*—a *totality* that has withdrawn already into the field in which it is deployed—there ceases to be anything at stake. The risk of questioning, the risk of constructing different strategies for inquiry, the risk of creating concepts is removed.

As such, the concept of technology introduces an issue, the

boundaries (and possibilities) of which always remain open. Technology presents, albeit in a sublime fashion, "thresholds" not yet determined, not yet fixed, not yet "present." (The concept of) Technology introduces the generation of concepts, where a concept presents an image (or images) of itself analogous to those images encountered in the cinema.[11] An image defines its own limits, marks out the boundaries of its field and the objects it represents, and orients the background of comprehension. In this way, technology is always an issue of itself, always present as and presented through images of itself. Comprehended in these terms, the identity of technology eludes apprehension; it is and will be, always, fragmented because the issue of technology is a self-dismantling process. The possibility of answering the question "What is technology?" is the possibility of the incessant creation of concepts. Thus, the anticipation of framing a peculiar kind of response that captures the essence of technology remains unfulfilled. Technology forces an ongoing exchange of questions, where responses can be translated into other questions and other ways of questioning.[12]

In a rather cryptic and telegraphic fashion, one might say that technology is its own model: a glimpse of itself given in artifacts, framed and fashioned according to the exigencies of a habitat and the visions of its inhabitants. The open issue of technology, what has been alluded to as the field of questioning "technology" designates, makes possible the *presentation of technology as such*. And yet, technology does not secure a foundation, a transcendental category or a necessary principle. It cannot be comprehended, if it can be comprehended at all, as Derrida remarks in *La Carte Postale*,[13] according to a Heideggerian *epoch* of Being where it would be submitted to a transcendental interrogation, a questioning that takes it "beyond every genre."[14] Instead, technology is a relay, a switch point within a series of relays or switching points "to mark that there is never anything but relays."[15] Its operations only leave the impression that technology (and/or science) is the "master switch" or the "switch box" controlling the exchange of relays.

5

Perhaps at this stage a more direct return to an etymological and lexical reflection on "technology" is appropriate. Indeed, as already indicated, an appeal to the etymological and lexical derivations indicates that there may be various ways for comprehending subtle differences in the use and deployment of a concept or a term. If one prefers, etymological reflection elicits a "recollection" (*anámnesis*), analogous to that discussed by Plato in the *Meno* (81c–d), of what is already presupposed and buried by the word's contemporary use. Here subtle differences initiated through a carefully nuanced articulation or inscription must incorporate the *artifice of usage* itself. Use creates. It orders, in each instance, a different concentration of significance or value, according to the philosophical, political, psychological, theological, and scientific topography in which the term is deployed. Because use involves repetition or reiteration, there is always the possibility of rupture: the conventions that have established the boundaries of a term or the limits of a concept are "by *essence* violable and precarious."[16] And this possibility of rupture is the possibility of *translation*: the creation of intermediary concepts and practices that can comprehend those that remain familiar and obvious. Translation is a form of displacement, the transfiguration of a habitat; it involves the transference of concepts and practices according to the artifice of *techné-logia*. Within such an etymological context, technology is bound to and merges with science.[17]

According to its rendering in *The Oxford English Dictionary*, "science" stems from the Latin *scientia* or "knowledge," a participle form of *scire* meaning "to know." According to its Latin roots (roots that can be linked to specific theological practices and designs), "science" designates the "state of 'knowing'" acquired by the study and mastery of a specific discipline, a particular branch of knowledge. Here "science"—the word and the *activity*—is used in a more restricted fashion. It is identified with "a connected body of demonstrated truths or with observed facts systematically classified and more or less colligated by being brought under general law." In its more recent determinations "science" develops "trustworthy methods for the discovery of new truth within its own domain." Thus, "science" is associated with a certain kind of "knowledge" or "learning," one that is distinct from art, as a craft, trade, or skill—*techné*.

But the juxtaposition of "science" and "*techné*" shows that the two activities were not always so dramatically opposed to one another. According to the Greek roots, *epistémé* ("true" knowledge) is concerned with "theoretic truth" whereas *techné* (art) is concerned with "methods for effecting certain truths." Where *epistémé* would be extended to denote "practical work which depends on the practical application of principles," *techné* would designate art that required "knowledge of traditional rules and skills acquired by habit." In some instances, in the context of Greek reflective

thought, science and art—*epistémé* and *techné*—are synonymous.
Theoretical knowledge requires practical application of principles;
art presupposes knowledge of rules. Moreover, there are moments
within this etymological context when "science" refers to a "craft,
trade, or occupation requiring trained skill."

In these latter determinations, *epistémé* and *techné* are already
interdependent. Knowledge of ideals, principles, and rules is knowledge derived from practice, from the need to satisfy a practical demand. In an analogous fashion, the engagement of ideas within a
specific context is functionally dependent on an awareness of concepts or rules derived from the habit of performance. One might
imagine this interplay in terms of what Gilbert Ryle calls the relation between "knowing-that" and "knowing-how."[18] Once again,
an appeal to etymology can assist in understanding this interplay.

As already noted, *scientia* is a participle form of *scire*. *Scire* has its
roots in *skei* which means "to cut" or "to split." Knowledge, then, is
understood as the *ability*, the *skill* "to separate one thing from
another," "to discern." In the Greek, such separation is related to
skhizein meaning "to split" into many parts, which is the root for
"schizo-" and "schism." Thus, the ability to discern differences, or
what Plato calls in *The Republic* a certain kind of "mindfulness"
(621c), is related to another Latin root *skel* which also means "to
cut" but which is more directly related to a concept developed later
in the Old Norse where "reason," "knowledge," "incisiveness" are
comprehended by *skil*, a precursor to our contemporary term
"skill." *Scire*, *skei*, and *skil* are derivations of *sek*, a basic form of
the Latin verb *secare* meaning "to cut." But here the separation, the
segregation of knowledge is not so acute. *Sek* is also "dissect,"
"exsect," "intersect," "notch," "resect," and "transect."

Thus, the pedestrian separation of "science" and "art," or "science" and "technology," is a function of the purposes—the theoretical
and practical ends—to which these words and activities are directed
within certain historical contexts. But what is more pertinent to the
purposes of this discussion is how the subtle differences in meaning,
the differences initiated through carefully nuanced articulation,
already presume and incorporate *the artifice of usage*, that is to say
"technology." Thus, "discernment," or what the Greeks understood as "scientific knowledge," presupposes some systematic
appreciation of "means" and "skill," means by which something is
created, such as the categorization of knowledge. The ability to discern is a condition for the possible separation of science and technology.

"Technology," as it is traced in *The Oxford English Dictionary*,

stems from the combination of two Greek terms *techné* and *logos*—
techné-logia. *Techné* designates art, craft, skill, a way or manner, a
means whereby something is created or contrived, devised by art.
Logos is rendered as discourse, account, definition, and proportion
(as an organizational principle). Combined, *techné-logia* is rendered
as a discourse or treatise on an art or arts; the systematic treatment
of the arts, sometimes the practical arts collectively. More recent
entries cite slight alterations and transformations of the term's usage: technology is identified with "weaving, cutting canoes, making
rude weapons, and in some place practicing a rude metallurgy," and
later in the nineteenth century it is linked to "applied science."

But now, what does the etymology reveal? Does it disclose anything more or less than what has been indicated already in the
normative definitions announced in *A History of Technology*?
Perhaps. Does the more direct appeal to the *word* pursued by an
etymological and lexical deliberation confirm the normative determinations? Does it affirm or negate the assertion that "technology
should mean"? I suggest that it does both *and* it does neither. The
reflection on the word "technology" and the history of its many uses
and determinations affirms that history: it is embedded in that history, it arises out of and against that background. An etymology of
"technology" reveals a history of questioning concerned with technology as it traces the derivations of its uses. By its very nature, the
reference to an etymology of "technology" is an attempt to trace
antecedent translations, not in an effort to imitate, to copy, or to
reclaim a lost and forgotten meaning. Instead, tracing the antecedent uses and renderings of "technology" (and "science") is pursued in order to sketch how translation, the reinscription of the term
and concept "technology," is a recognition of the difficulty, if not
the impossibility, of defining "technology" and submitting such a
determination to a specific type of interrogation. Translation affirms
antecedent renderings, but in doing so it negates or denies their
permanence. Translation is not identical to displacement. Indeed,
there are no identities here; there is no comprehension of technology. There are only appearances of technology in the form of
something—a concept, a name, a practice—other than itself. But
as displacement, translation entails the transfiguration and transformation of what is, of what is said to be . . . Thus, however technology is identified, however it is to be accounted for and controlled, whatever account of technology is presented, that cannot be the
last word. The discourse on *techné* prohibits finality.

To be sure, an etymology provides a sketch; it is already a trace
and a tracing of antecedent translations. Every retracing of this

sketch constitutes a translation, another sketch. And in this way it frames a certain image of technology—one that either confirms an already dominant conception or one that forces a refashioning of an old conception. Herein lies the difficulty of definition and translation. All the predicates, all the defining concepts, all the lexical significations, that at once seem to fall under a particular rubric, are conditioned by and products of technology. Like any other term or concept, the value of "technology" is determined by its use in a network of possible substitutions, in a topography marked by the ambiguity of its constitution. All too often, and all too carelessly, this topography is called a "context." It is forgotten that the context in which a concept takes shape is inexorably composed of ethical, political, artistic, literary, psychological, sociological, and scientific interests, and is fabricated in and through the use of that concept.

For the purpose of this experiment in rethinking the concept of technology, technology is of "philosophical" interest because it can be inserted into a chain of other equally overdetermined terms and concepts. It displaces some terms and comes to be determined by the other terms in the chain, such as art, artifice, creation, discourse, craft, cunning, dissimulation, science, skill. As the etymologies of "science" and "technology" indicate, this list by definition can never be closed. It marks off the medium or the means for revealing what John Dewey calls a "sequential bond" with a "foreseen" future.[19] Even though the future is "foreseen," already envisioned, it remains *unpresentable today*; this is an effect of the medium in which it is to be presented—technology. In the present or current uses of the substitutions that supply this list of terms, then, the future is designed, marked out, and created; yet it remains deferred, the product of technological artifice, the fable of so many discourses.

Within this context, technology presents its "usefulness," its legitimacy—certain ways of doing, certain artifacts and techniques to be employed in specific fashions for definite purposes. But the determination of "usefulness" or "practicality," all of what has come to be called "applied," is conditioned by its being a product of use, a product of the performance of *and* the discourse on artifice, *techné-logia*.

6

Perhaps the hypothesis that opens this inquiry into technology's "significations" should be rewritten. Perhaps the statement should read "the *operations* of technology are presupposed by and legitimate all questioning"? Why? What difference does the addition of the term "operations" make in the force of the announcement, in its attempt to force a rethinking of technology or frame a line of questioning? It provides another means for indicating how the operations of technology (art, artifice, creation, translation, mediacy, framing, and so on) are always already at work in whatever investigations or projects we set out to complete. The operations of technology mediate our practical and intellectual activities, and in their performance our comprehension of technology is mediated— altered, designed, displaced, resituated, constituted.

Moreover, to speak of "the operations of technology" is to draw attention once more to the assemblage of terms already incorporated into the discussion that indicate the "operations." Technology is not a "thing," even though there are things that go by the name "technology." But even if it is conceived as a "thing" or a tool, technology becomes what it is, or at least what it is said to be, always in the possible realization of possibilities. The operations of technology are "connective operations."[20] The tool is significant *not* in its immediacy as a tool, but in its *mediacy, as a medium* for knowing the future in the present. If technology marks off a medium for revealing a "sequential bond" with a "foreseen" future, then the artifacts and objects of technology do not designate a place in which the coming-to-be of the future will occur. The significance of technological artifice always lies elsewhere, always assuming another form— in the connective operations and bonds that constitute our habitat.

Through a series of all too abrupt and fleeting reflections, I have tried to demonstrate that the concepts and practices incorporated by the word "technology" cannot be comprehended within the limits of analytical or historical narratives that trace the development and use of implements, and the circumstances that lead to their formation. And yet, it is only in and through such narratives that technology takes on any form whatsoever and can thus be comprehended. It is difficult, if not impossible, for these (or any) narratives to take into account, as well as give an account of, their involvement in the scene they describe. Instead, I have tried to show that there is an active and internal alliance among

1. concepts (here the concepts of technology, science, and art),
2. the operations of artifice that condition the creation of concepts,
3. the conceptual effects and sensible artifacts through which we frame the future, and
4. the domains in which these concepts and artifacts are enacted and engaged.

The concept of technology frames our concerns with modes of habitation. It "inhabits" and "occasions" them, whatever they may be, however they may appear.[21] Technology carries us, transports us, transforms and transfigures our dwellings; displaces us, and translates (or makes possible) the images of the future that inform the "present." If concepts can be likened to the images encountered in the cinema, then, in their projection, not only do these images define their own limits, mark out the boundaries of their field, and orient the background against which we comprehend the image, they do so because they identify what Gilles Deleuze calls an "out-of-field" (le hors-champ).[22] The "out-of-field" is not a negation; it "refers to what is neither seen nor understood, but is nevertheless perfectly present."[23] It refers to the unpresentable, that which cannot "fit" within the limits of a frame. The "out-of-field" refers to that which has withdrawn into the structures of thought, action, and discourse. With respect to technology then, it is its presence apart from its sensible representation—that which "does not belong to the order of the visible"—that continually needs to be rethought.[24] The presentation of technology itself as the issue, as the untranslatable condition and medium of possibilities continually needs to be rethought.

Whatever the concept of technology, it fails to comprehend the entire field of its possibilities; it cannot comprehend the topography in which it is enacted or inscribed, even though technology refers to what is both included in its framework and what is out-of-field. Whatever the concept of technology, it cannot comprehend the "technology," the artifice, the discourse, the techné-logia that structures it and makes it possible. Just as our concepts or projected images of technology, our uses of the term, frame a parcel of this topography for which borders are established, they are part of that parcel. Concepts localize, isolate possibilities, and thus translate the complex scene of habitation into a more condensed field of vision. Concepts and practices overflow their supposed boundaries: images at one level of representation always intrude upon and determine the facts and images at another level of representation. Concepts actively and internally constitute an assemblage of possibilities, an assemblage of images always transforming itself, translating over and over again our concern with the conditions under which we would like to live—our habitat. Within this assemblage, technology is one thread, among many, traversing conceptual and practical boundaries, making possible the articulation of one discourse within the domains of another. Does this not indicate the operations of technology? Is this not the issue of technology—the creation and weaving of intermediary concepts and practices?

NOTES

1. Edgar Allan Poe, *Collected Works*, vol. 3, ed. Thomas Ollive Mabbitt (Cambridge: Harvard University Press, 1978), p. 985.
2. Friedrich Nietzsche, *The Birth of Tragedy*, trans. Walter Kaufmann (New York: Vintage Books, 1967); see especially Nietzsche's second preface entitled "Attempt at a Self-Criticism [*Versuch einer Selbstkritik*]," pp. 17–27.
3. Cf. Don Ihde, *Existential Technics* (Albany: State University Press of New York, 1983), especially "The Historical-Ontological Priority of Technology over Science," pp. 25. Ihde claims that "technology, particularly in its more recent developments, is the *condition of the possibility of science.*" By advancing the claim that "technology is presupposed in and legitimates all questioning," I am not in disagreement with the fundamental thesis Ihde presents. However, I am interested in articulating a different set of themes or issues that are not bound to assertions regarding any kind of priority—ontological, historical, epistemological, and so on—of technology over science. Instead, I am interested in articulating *those conditions that make possible such announcements*.
4. Charles Singer, E. J. Holmyard, and A. R. Hall, *A History of Technology* (Oxford: The Clarendon Press, 1954–78), 8 vols.
5. Ibid., especially V. Gordon Childe, "Early Forms of Society," vol. 1, pp. 38–57. See also Melvin Kranzberg and Carroll W. Pursell, Jr., "The Importance of Technology in Human Affairs," in *Technology in Western Civilization: The Emergence of Modern Industrial Society. Earliest Times to 1900*, vol. 1, ed. Melvin Kranzberg and Carroll W. Pursell, Jr. (Oxford: Oxford University Press, 1967), pp. 4–6. Kranzberg and Pursell note that any definition or conception of technology must include some reference to that which is "man-made" and the effects of that production on the environment or "system" in which humans live and work. They write: "Technology . . . is much more than tools and artifacts, machines and processes. It deals with *human work*, with man's attempts to satisfy his wants by human action on physical objects" (p. 6).
6. Cf. Kranzberg and Pursell, "The Importance of Technology in Human Affairs," in *Technology in Western Civilization*, p. 7–11.
7. Cf. Ludwig Wittgenstein, *Philosophical Investigations*, trans. G. E. M. Anscombe (New York: Macmillan, 1968), pt. 1, sec. 68–72.
   The question of "what counts as technology and what no longer does" is pertinent to a remark found in Kranzberg and Pursell's "The Importance of Technology in Human Affairs." Charles Singer defines the subject of technology, and the history of technology, as "'how things are commonly done or made . . . [and] what things are done or made.'" Kranzberg and Purcell write that "Such a definition is so broad and loose that it encompasses many items that scarcely can be considered as technology. For example, the passage of laws is something which is 'done,' but the history of law certainly is not the history of technology" (*Technology in Western Civilization*, p. 5). Even if these two histories are not identical, this is not to say they do not depend on one another for the order they bring to a very diverse and complex set of phenomena. Is a history of technology possible apart from a history of law, or apart from the *use* made of law to insure the development of certain "technologies"? Likewise, is a history of law possible apart from a history of technology, apart from the *use* made of technology in the legitimation of certain judicial practices?
8. "Preface," by Singer, Holmyard, and Hall, in *History of Technology*, 1. vii.
9. Childe, "Early Forms of Society,," p. 38; cf. Kranzberg and Pursell, "The Importance of Technology in Human Affaris," p. 6. "We must use the term 'wants' instead of 'needs,' for human wants go far beyond human needs, especially those

116   GAYLE L. ORMISTON

basic needs of food, clothing, and shelter. Technology administers to these, of course, but it also helps man to get what he wants, including play, leisure, and better and more commodious dwellings."

10. "Preface," in *History of Technology*, 1: vi–vii.

11. See Gilles Deleuze, *Cinema I: L'Image-Mouvement* (Paris: Les Éditions Minuit, 1983), pp. 46–61 and 243–89; see the English translation by Hugh Tomlinson and Barbara Habberjam, *Cinema 1: The Movement Image* (Minneapolis: University of Minnesota Press, 1986), pp. 29–40 and 178–215.

12. Cf. Martin Heidegger, "The Question Concerning Technology," in *The Question Concerning Technology and Other Essays*, trans. William Lovitt (New York: Harper and Row, 1977), pp. 3–4.

13. Jacques Derrida, *La Carte Postale: De Socrates à Freud et au-delà* (Paris: Flammarion, 1983) p. 206; *The Post Card: From Socrates to Freud and Beyond*, trans. Alan Bass (Chicago: University of Chicago Press, 1987) pp. 191–92. See also Derrida's remarks on the relation between "writing" and "technics" in *Of Grammatology*, trans. Gayatri Chakravorty Spivak (Baltimore: Johns Hopkins University Press, 1976), p. 8. Derrida writes: "I believe . . . that a certain sort of question about the meaning and origin of writing precedes, *or at least merges with*, a certain type of question about the meaning and origin of technics. That is why the notion of technique can never simply clarify the notion of writing" (emphasis added).

14. Derrida, *La Carte Postale*, p. 206; *The Post Card*, pp. 191–92.

15. Ibid.

16. Jacques Derrida, "LIMITED INC a b c . . . ," trans. Samuel Weber, *Glyph 2: John Hopkins Textual Studies* (1977): 250.

17. A more elaborate discussion of these etymologies and their interrelationships is provided in Gayle L. Ormiston and Raphael Sassower, *Narrative Experiments: The Discursive Authority of Science and Technology* (Minneapolis: University of Minnesota Press, 1989), chap. 1, "The Interplay of Science and Technology: An Introduction."

18. Gilbert Ryle, *The Concept of Mind* (New York: Barnes and Noble, 1949) p. 54. Ryle maintains the separation of these designations. I refer to "knowing-how" and "knowing-that" as convenient terms for designating two sides of a very complex, reciprocal relation between kinds of activity, ways of talking, modes of understanding. The use of one form (or each form, as it were) of "knowing"—"knowing how"—signifies the mediate presence of the other form—"knowing that."

19. John Dewey, *Experience and Nature* (New York: Dover Publications, 1958), pp. 122–23 and 152; cf. Martin Heidegger, "The Origin of the Work of Art," in *Poetry, Language, Thought*, trans. Albert Hofstadter (New York: Harper and Row, 1971), p. 59. Heidegger writes: "*techné* signifies neither craft nor art, and not at all the technical in our present-day sense; it never means a kind of practical performance. . . . The word *techné* denotes rather a mode of knowing. To know means to have seen, in the widest sense of seeing, which means to apprehend what is present as such."

20. John Dewey, *The Quest for Certainty: A Study of the Relation between Knowledge and Action* (New York: G. P. Putnam's Sons, 1929), p. 268.

21. Cf. Derrida, *La Carte Postale*, p. 207; *The Post Card*, p. 192.

22. Deleuze, *Cinema I*, p. 28; *Cinema 1*, pp. 15–16.

23. Ibid., p. 28; p. 16.

24. Ibid., p. 30; p. 17.

The Autonomy of Technology

JOSEPH C. PITT

> Technique is autonomous with respect to economics and politics.
> . . . Technique elicits and conditions social, political, and
> economic change. It is the prime mover of all the rest, in spite of
> any appearance to the contrary and in spite of human pride,
> which pretends that man's philosophical theories are still deter-
> mining influences and man's political regimes decisive factors in
> technical evolution
>
> Jacques Ellul, *The Technological Society* ·

Some features of Ellul's views are suggestive—for example, his belief that technology has a way of precipitating certain economic results. However, I find less convincing a central theme in Ellul's
work, the autonomous status of technology and its impact on society. In this essay, I argue that speaking of the "autonomy of technology" commits one to a fundamentally wrong way of addressing
issues in technology studies.

It has become commonplace to think of technology as an independent kind of "thing," that is to reify technology and then attribute
causal powers to it. However, it is not clear that reifying technology
moves the discussion anywhere except down blind alleys. Thus, for
example, it is only after the first move toward reification that we
hear about such things as the "threat" of technology taking over our
lives. Likewise, reification leads to misleading talk about technology
being the handmaiden of science, or some variant on that theme. In
other words, reification makes talk about autonomy possible. Yet,
in my opinion, there is no useful sense in which we can conceive of
technology as autonomous.

Originally published in *Technology and Responsibility, Philosophy and Technology* series, vol. 3, edited by Paul T. Durbin (Dordrecht: D. Reidel Publishing,
1987), pp. 99–114.

Let me stress the "useful" here. Technology is such an integral part of our society and culture that unless we ferret out the way in which it actually is embedded in our world, we may fall victim to a kind of intellectual hysteria that makes successful dealings with the real world impossible. The first step we need to take to avoid this danger is to clarify the kinds of issues that reasonably can be addressed.

NONISSUES

There are at least two cases of talk about the autonomy of technology that are nonstarters. Although they are popular topics of discussion, if considered carefully, we will see that they are irrelevant since the kind of autonomy involved is trivial.

In the first case, it is claimed that technology is autonomous when the inventor of a technology, once the technology is made available, loses control of his invention. For example, once digital computers entered the public domain, it was impossible for anyone to call them back. The rapid sophistication in their complexity and the all-pervasiveness of their employment in society was impossible to stop. Surely, the story goes, this is a case of autonomous technology.

Well, yes and no. Yes, it is autonomous, if by that is meant only that the inventor alone can no longer control the development of the technology. But, this is a trivial sense of "autonomy" because it is true of all aspects of our society. Once in the public domain each item is beyond the control of its inventor in some sense or other. But that does not make the item autonomous. Its further development is a direct function of how people employ it and extend it. To the extent that people are necessarily involved in that process, the invention cannot be autonomous.

In the second trivial case of autonomous technology, it is claimed that because the inventor of a technology has failed to see the consequences of employing a given technology in a certain way, the technology is autonomous. Yet no one can foresee all the consequences of any move. That fact, however, does not entail that once some action is taken, the consequences of that action are autonomous. Although the full consequences of the introduction of the automobile were not anticipated by Henry Ford, it does not follow that those consequences were due to the autonomy of the automobile.

The key to understanding this second point lies in realizing that once a piece of technology leaves the hands of its inventor it also leaves behind the situation in which the actions of only one person can affect its development and employment. Once in the public domain the diffusion of a technology will be the result of community decisions; these kinds of decisions are at best the results of compromises. That there is no logical order to the patterns these decisions take should come as no surprise. Compromise is a function of a variety of factors and it is impossible to tell in advance which of them will be prominent in any given situation. It may be that this lack of absolute predictability is what provides the illusion of the autonomy of the technology. But the fact that the role a technology takes on in a society is a function of community decisions, which are at best compromises, does not entail that the technology is autonomous. It would appear quite the contrary, that the technology is subjected to the kind of buffeting and manipulation this process involves.

Thus, arguments from the eventual lack of control of the inventor and the failure to foresee all the consequences fail to secure the case for the "autonomy" of technology. But there are also other arguments we need to consider.

THE PROCESS OF TECHNOLOGY

Well-intentioned writers and critics have commented on various aspects of technology that they see as raising the possibility of autonomy and with it the specter of some version of the apocalypse. One of the best examples of the kind of worry expressed by these authors can be found in John McDermott's essay review of Mesthene's *Technological Change*, "Technology: The Opiate of the Intellectuals."[1] There McDermott speaks of a kind of momentum certain technologies appear to take on when employed in given ways, thereby providing the appearance of autonomy. Consider the following McDermottian scenario: a growing retail company has just hired an up-to-date accountant to manage the books, which over time had deteriorated to the point of chaos. Our accountant is a bright young modern man and, given the size of the company and its projected growth, argues persuasively that in the long run it will be cheaper and more efficient to buy a couple of computers than to hire an entire accounting staff. He produces a report showing the projected costs of people versus machines, projecting only for the long run the cost of benefits and retirement for the people and maintenance for the machines. He wins his case and the computers are purchased. But once computers are introduced, air conditioning is not far behind because the computers need a cool environment to function maximally. But, our fictional tale continues, air conditioning requires that the entire building be redesigned to handle air flow and pressure. Finally, our storyteller says with a knowing look, the president of the company is driven to complain: "How did we get into this fix? The old building is perfectly good, we don't really need air conditioning; since we introduced those machines it looks like I have lost control!"

This, I submit is a typical story—one even sometimes told—and perhaps representing a situation often experienced. But, just because the story is told and some people may interpret their experience in this fashion, it does not follow that this is the way things are. What this tale allows us to see is that despite the fact that machines play a prominent role in the unfolding sequences of events, the major overlooked fact is that people often tend to forget the reasons for which they introduced a certain kind of tool or procedure and, instead of taking time to assess critically the impact of making further accommodations to the technology, possibly concluding that it may be time to reexamine the whole situation, simply "go with the flow" and take the course of least resistance. However, from the fact that people tend to react to the presence of technologies in certain ways, perhaps to accommodate the technology at first rather than to replace it with another, it does not follow that the technology is autonomous.

A basic point that we sometimes tend to forget is that there is no getting rid of technology *simpliciter*. Man or woman in the world is man or woman using tools of wide variety and complexity (hammers, automobiles, governments). The tools we invent to help us achieve our goals are essential, perhaps even to the concept of man or woman. We cannot remove technology, all of it, and continue without. When we introduce an implement or a system, it is to help us achieve a given goal. If we find that device produces results or side effects that are in conflict with other goals and/or values we hold, we may replace it or modify it (à la Langdon Winner).[2] Whichever, technology remains with us; it is a part of how we go about making our way in the world. What McDermott overlooked (when he spoke of how technologies become so ingrained in our procedures that in accommodating the requirements of the technology we lose our independence of action) was that it is the perception, or lack of it, that people have of the usefulness of a technology in the pursuit of a goal that determines the extent to which they are willing to make concessions. They may also tend to lose sight of the goal that first guided their action and, therefore, they react blindly to the circumstances with which they are faced. But that is not to say that the technology has taken over. For nothing in principle rules out later modifications and, if necessary, replacements. What is required is that the individuals involved keep their objectives in mind. And indeed, that is what technological assessment is about. But before we can see how the individual functions here, we ought to look at how the picture has been misconceived from the beginning.

RECHARACTERIZING TECHNOLOGY

There appear to be at least two contexts in which the autonomy of technology has been an issue: in its *use* and in its *development*. In the first case, as noted previously, it is sometimes alleged that once brought into play, certain technologies take on a life of their own that impels them into our affairs without human guidance or control. In the second case, there is the familiar claim that technology is merely applied science, or the results of scientific theorizing made practical, and that, in point of fact, technology does not arise independently of science nor does it have an autonomous origin and history. If we take a quick look at the first point we can note that an examination of the possibility of technology running away from human control reveals that it is a worry without merit. Our accounting tale above ought to help make that point. But more is needed than telling stories. To defuse fears about uncontrolled autonomous technology requires recharacterizing the concept of technology. To that end, I am urging that rather than think of technology only as products of certain kinds, that we think of technology as a process that includes: deliberation and policy formation, implementation and use of tools and systems, and a feedback mechanism leading to information updating and assessment procedures.[3] Once technology is viewed as this process whereby human beings engage the world, including the time it takes to assess the impact of their efforts and to readjust their strategies and procedures in the light of what their efforts have revealed, fear of an autonomous technology degenerates into pathology. For the question no longer concerns technology, but rather humanity's willingness to take the necessary steps to bring about desired changes. This is not to say that bringing about these changes is easy—but only that it is not out of our hands to do so. I will return to this issue at the end.

If I am correct and the idea of autonomous technology at loose in the world without human control is merely a function of misconceiving the nature of the beast, then what about the second version of

autonomous technology mentioned—the independence of science
and technology, or, minimally, the independence of science from
technology in the development of science? The basic idea here
is that technologies cannot develop independently of scientific
theories, that technology is merely applied science. The subsidiary
idea is that science, however, does not need technology for its own
growth. Both notions are false. But, except for careful historians of
science and technology such as Hall, Calggett, I. B. Cohen, Crom-
bie, Drake, Dumas, and Price, the popular account has it otherwise.
Furthermore, most philosophy of science also proceeds, to its
detriment, as if the above claims were true. To my mind it is far
more important to clarify these issues than to try and contend with
reactionary nay-sayers and the four horsemen. Furthermore, the
questions these issues raise also can be answered by invoking the
same analysis of technology applied previously (to questions about
autonomous technology allegedly taking over human affairs), and
coupling that analysis to some history of science and technology.
If technology is a process that includes the implementation of certain strageties to solve problems, and an assessment of the success
or failure of those strategies followed by an integration of the information gleaned from the assessment into the general body of
knowledge used to make further plans upon which action is based,
then any attempt to see technology as merely a product of science
with no further input into the science not only does injustice to the
history of technology, but also fails to understand human action.

COMMON SENSE

Phrased as I have put it, technology represents common sense; it
is how humans gain experience and what it means for men and
women to act on their experience. Nor should that come as a surprise. Since, if we acknowledge that the concept of a tool lies at the
commonsense heart of technology, and if we accept the rather obvious point that not all tools are physical tools (i.e., that there are
conceptual tools, social tools, economic tools, etc.), then it is not
difficult to agree that knowledge is a tool, and if knowledge is constantly being updated, the tool is constantly being honed. In other
words, if science produces knowledge, then the knowledge science
produces is constantly being upgraded and changed by virtue of the
impact of technology on the efforts of science to discover more and
more about the world. Quite aside from the resolution of the question of the independence or interdependence of technology and sci-

ence, if science produces knowledge and that knowledge is sometimes used to develop tools that are used in the world, then what those tools produce should generate a form of empirical knowledge that ought to bear back on the original knowledge that produced the technology. In addition, it follows that technology is also constantly being changed in the face of these developments, and that is as it should be. The bottom line is that, on this account, once a relation between a science and some technology is established, neither can lay further claims to autonomy—the interdependence is an essential aspect of the process of science itself. But this point of view cannot be established only by *a priori* argument. We need to look at what actually goes on; and I have selected a historical case study to illustrate my points. This is not to say that the analysis of one historical example will settle the issue, but it should help to clarify some matters.

Indeed, the case of Galileo and the telescope ought to help exhibit just the issues I see relevant to sorting out some of the confusions surrounding the interrelations between the development of science and use of technology. Furthermore, there is a bit of a punch line. The general thesis, as already expressed, is that science and technology, where they interact at all, are mutually nurturing. This, of course, has to be bracketed by virtue of a caveat, to wit, in point to fact some technologies are science independent, for example, the roads of Rome (which is not to say autonomous, since those technologies were responses to needs and goals also, just not the needs and goals of some scientific theory) and some science generates no technology, for example, Aristotelian biology. But something of a paradox emerges. For the history of science is the history of failed theories. But the failure of the theories most often does not force a discarding of whatever technology that theory generated or was involved with, nor does the failure of the theory force the abandoning of the technology if a technology was responsible for that theory. Oversimplified; sciences come and go, but their technologies remain. But oversimplification got us into trouble at the start, so a more accurate claim would be: scientific theories come and go, but some technologies with which they are in one way or another associated remain.

But there is one sense in which this point becomes somewhat problematic. That is, if, as I put it earlier, technology is an integral part of science and partially responsible for changing the science, then the failure of the particular theories could be construed as a failure of the technology involved as well. To respond to this requires that we emphasize that technology is a process of policy

formation, implement/system implementation, assessment and updating that functions at a variety of levels and with varying degrees
of significance for technologies further up and down the line—for
example, the failure of a rocket may mean the failure to meet a
certain deadline, but it does not spell disaster for the entire shuttle
project. Goal achieving activities are nested within one another and,
as we shall see, as a matter of historical and physical accident the
nesting will have different degrees of importance depending on the
case.

GALILEO AND THE TELESCOPE

To illustrate some of the notions introduced here let us turn to an
examination of the development of the telescope by Galileo and its
effect on some of the theoretical problems he faced in his efforts to
show that Copernicus's theory was worthy of serious scientific consideration. As we shall see, the story is not a simple one and the
issues take on an increasing degree of complexity as the tale proceeds.

To begin with, we need to be perfectly clear that Galileo did not
begin his work on the telescope in order to prove anything about
Copernicus. The full story of how Galileo came to construct his first
telescope is clearly and succinctly put forth by Drake in his *Galileo
at Work*.[4] There, quoting from a number of Galileo's letters and
published works, Drake shows that Galileo was first drawn to the
idea of constructing a telescope out of financial need. In July 1609
Galileo was in poor health and, as always, if not nearly broke at
least bothered by his lack of money. Having heard of the telescope,
Galileo claims to have thought out the principles on which it worked
by himself, "my basis being the theory of refraction."[5] Drake acknowledges that there was no theory of refraction at the time, but
excuses Galileo's claim on the grounds that this was not the first
time that Galileo arrived at a correct result by reasoning from false
premises. (Historians of the logic of discovery, take note!) Once
having reconstructed the telescope, Galileo writes: "Now having
known how useful this would be for maritime as well as land affairs,
and seeing it desired by the Venetian government, I resolved on the
25th of this month [August] to appear in the College and make a
free gift of it to his Lordship."[6] The result of this gift was the offer of
a lifetime appointment with a nice salary increase from 520 to 1000
florins per year. What was unclear at the time, and later became the
source of major annoyance on Galileo's part, was that along with

the stipend came the provisions that there was also to be no increase for life! So he reinitated his efforts, eventually successful, to return to Florence.

Now there are some problems here that need not delay us, but they ought to be mentioned in passing. How Galileo managed to reconstruct the telescope from just having heard reports of its existence in Holland remains something of a mystery. Galileo provides us with this own account of the reasoning he followed; but, as Drake notes, his description has been ridiculed by historians because, despite the fact that the telescope he constructed worked, he did not think the project through correctly. Nevertheless, Drake's observation that "the historical question of discovery (or in this case, rediscovery) relates to results, not to rigorous logic"[7] seems to the point. Despite the fact that a telescope using two convex lens can be made to exceed the power of one using a convex and a concave lens, the fact of the matter is that Galileo's telescope worked. On the other hand, this point about faulty reasoning leading to good results seems to tie into the paradoxical way in which technologies (thought of as artifacts of varying degrees of complexity and abstractness) emerge and remain with us. But more of this later.

We can now turn to the question of the impact of the telescope on Galileo's work. As he reports it, Galileo first turned his original eight-power telescope toward the moon in the presence of the Grand Duke of Florence, Cosimo. He and Cosimo apparently discussed the mountainous nature of the surface of the moon and shortly after his return to Padua in late 1609, Galileo built a twenty-power telescope, apparently to confirm his original observations of the moon. He did so and then wrote to the Grand Duke's secretary to announce his results. So far then, Galileo has constructed the telescope for profit and is continuing to use it to advance his own position by courting Cosimo.

Galileo, never retiring about his work, continued to use the telescope and make his new discoveries known through letters to close friends. Consequently, he also began to attract attention. But, others such as Clavius now also had access to telescopes. (It appears that reinvention was not uniquely Galileo's strength.) That meant Galileo had to put his results before the public in order to establish his priority of discovery. Therefore, in March 1610 Galileo published *The Starry Messenger*, reporting his lunar observations as well as accounts of the Medicean stars and the hitherto unobserved density of the heavens. And now the fun begins. For these reports essentially challenge one of the fundamental assumptions of the Aristotelian theory of the nature of the heavenly sphere, its perfection and immutability. While the rotation of the Medicean stars around Jupiter can be shown to be compatible with both the Copernican and the Tyconian mathematical astronomies, it conflicts with the philosophical and metaphysical view that demands that the planets be carried about a stationary earth embedded in crystalline spheres. And to be clear about the way the battle lines were drawn, remember that Galileo's major opposition came primarily from the philosophers, not from the protoscientists and other astronomers of this time.

The consequences of Galileo's telescopic observations were far more reaching that even Copernicus's mathematical model. For the problems Copernicus set were problems in astronomical physics and as such had to do with meeting the observational restraints represented by detailed records of celestial activity. Galileo's results, however, and his further arguments concerning the lack of an absolute break between terrestrial and celestial phenomena, maintaining as he did the similarities between the moon and the earth, forced the philosophers to the wall. It was the philosopher's theories that were being challenged when the immutability of the heavens was confronted with the Medicean stars, the phases of Venus, sunspots, and new comets. One might conclude, then, that this represented something akin to a Kuhnian gestalt switch.

A lot of silliness has been written about the extent to which Kuhn's paradigm shifts and their purported likeness to gestalt switches actually committed someone who experiences one to seeing a new and completely different world. But to see mountains on the moon in a universe in which celestial bodies are supposed to be perfectly smooth comes pretty close to making sense of what this extreme interpretation of Kuhn might mean. Prior to the introduction of the telescope, observations of the heavens, aside from providing inspiration for poets and lovers, were limited to supporting efforts to plot the movements of the planets against the rotation of the heavenly sphere. Furthermore, metaphysical considerations derived from Aristotle interfered with the conceptual possibility of learning much more, given the absence of alternatives of equal or greater explanatory coherence. The one universally accepted tool that was employed in astronomical calculation was geometry and its use was not predicated on any claims of realism for the mathematical models that were developed, another point derived from Aristotelian methodology. The acceptable problem for mathematical astronomy was to plot the relative positions of various celestial phenomena, not to try to explain them. Nor were astronomers expected to astound the world with new revelations about the population of the heavens since that was assumed to be fixed and perfect. So, whatever else astronomers were to do, it was not to discover new facts—there were not supposed to be any.

But the telescope revealed that there were new facts. And for Galileo this entailed that some way had to be found to accommodate them. Furthermore, to make the new telescopic findings acceptable, Galileo had to do more than merely let people look and see for themselves. The strategy he adopted was to link the telescopic data to something already secure in the minds of the community: geometry. This, however, was not as simple as it sounds. He had to build a case for extending geometry as a tool for physics, thereby releasing it from the restrictions under which it labored when used only as a modeling device for descriptive astronomy. In other words, Galileo had to advance the case of Archimedean mechanics. To this end he was forced to do two different things: (1) emphasize rigor in proof—extolling the virtues of geometry and decrying the lack of demonstrations by his opposition, and (2) deemphasize the appeal to causes in providing explanations of physical phenomena (since abandoning the Aristotelian universe entailed abandoning the metaphysics of causes and teleology without which the physics was empty).

GEOMETRY AS A TECHNOLOGY

This is not the place to detail the actual way in which Galileo employed geometry to radicalize the notions of proof, explanation, and evidence.[8] Suffice it to say that he did and that it met with mixed success. The general maneuver was to begin by considering a problem of terrestrial physics, proceed to "draw a little picture," analyze the picture using the principles of euclidean geometry and (1) interpret the geometric proof in terrestrial terms, just as a logical positivist would interpret an axiomatic system via a "neutral" observation language, and then (2) extend the terrestrial interpretation to celestial phenomena. This is how he proceeded with his account of mountains on the moon, namely by establishing an analogy with terrestrial mountains. This process took place in stages. He first subjected the terrestrial phenomena to geometric analysis and then he extended that analysis to the features of the moon. Not all of Galileo's efforts at explanation using this method succeeded; consider, for example, his account of the tides. Nevertheless, the central role of geometry cannot be denied.

While Galileo had flirted with geometry for most of his career, it was not until he was forced to support publicly his more novel observations and hypotheses that we find in his writings the beginnings of what eventually was to become a very sophisticated methodological process. This procedure is most clearly evident in his last two works, the *Dialogue on the Two Chief World Systems* and his *Discourses on Two New Sciences*. But in the end the more *geometrico* as employed by Galileo, or to put it slightly differently, Galilean science, dies with Galileo. No one significant carried on his research program using his methods. Whatever impetus he gives to mathematics in science, his mathematics, geometry, very quickly gives way to Newton's calculus and the mathematics of the modern era.

Now it seems to me that Galileo's use of geometry was as much the employing of a technology as was his use of the telescope. Furthermore, it represents the first major step toward the mathematization of what today we would call science. This much is commonplace. The interesting part comes in two sections: (1) the telescope was a new technology, whose introduction for initially nonscientific reasons (i.e., money) was in fact science independent, for its invention by the Dutch was theory independent.[9] In many ways, the use of this new technology by Galileo can be held responsible for the extension of the more geometrico as a radical method of supporting knowledge claims. (2) Geometry was also theory independent. But, unlike the telescope, this was a very old technology. It was called upon to rescue, as it were, the new technology. It is a very different kind of technology from the telescope, being a method for providing justifications, or proofs, of abstract conclusions regarding spatial relations, not a physical thing. Furthermore, despite the fact that this old technology was required to establish the viability of the new; the old was soon to become obsolete with respect to the justificatory role it was to play in science. That it was to be replaced also had nothing to do with any significant relation between the telescope and the development of the theory Newton outlined in his *Principia*. In other words, the telescope itself had little direct bearing on the development of the calculus, and yet it was the calculus that superseded geometry (but did not completely eliminate it) as the mathematical basis for scientific proof.

TECHNOLOGY AND THE DYNAMICS OF CHANGE: AUTONOMY SOCIALIZED

If we try to sort it all out, the results are uncomfortable for standard views of technology and the growth of knowledge. The two technologies remain, the two sciences have been replaced. Furthermore, in one of the truly nice bits of irony that history reveals, one of the discarded technologies, geometry, after being replaced by a different kind of mathematical system for justificatory functions, experiences a resurrection in the nineteenth century and ends up playing a crucial role (but not justificatory role) in the development of yet another physics, having been modified and expanded in the process.

Where is the autonomy here? Both Galileo's physics and the telescope, while capable of being viewed as independent products of one individual's creative energy, can also be seen performing an intricate *pas de deux* of motivation and justification when the process of inquiry is examined. It is getting difficult to determine which view ought to take priority. A resolution to the problem might be found if we stop looking at the history and examine the concept of "autonomy" itself.

If we define "autonomy" as "free from influence in both its development and its use," then technology cannot be autonomous because it is inherently something used to accomplish specific goals. But what happens if we try to define technology so as to deliberately allow technology to have an impact on us as well as on our environment? Are we then committed to the view that, given a technology in use, there emerges from its use a self propagating process outside the control of humankind? If (1) technology is a product, and (2) unless we add some additional properties to technology beyond it being a thing we manipulate, then (3) there is no reason why we should even begin to think of technology as not within our control.

In other words, we can talk of Galileo being forced to employ geometry and develop novel methods of justification in order to defend his telescopic discoveries, but what sense does "forced" carry here? The telescope did not with logical necessity precipitate him headlong into battle. Much of what Galileo did to defend his claims and insure his priority of discovery was the product of his flamboyant personality. This was a man who loved fights and being in the public eye. How these features of Galileo's personality can be factored into the technology so as the make it appear that the tool itself is responsible for the action of the man is beyond me. Given the technology we can plot its history. What that history amounts to is how it is used. How it is used is a complicated process, for it can entail more than intentional application of a device. "Use" may also mean "rely on" and it may be the case that what we rely on we take for granted, never giving thought to the cost. But this does not thereby entail that in the absence of human deliberation, the tool by default acquires intentionality and along with it control of human affairs.

An alternative would be to endorse the idea that both the telescope and geometry used Galileo. This suggests a scientific fiction scenario in which as soon as any technology is touched by a man or woman it "takes over" that indiviual. In the case of populations adopting constitutions that establish governments, then all freedom of human action is lost since the government—thing reified—"takes over." Surely this amounts to a reductio. For the tool used to adopt the government is reason. Is reason too going to be something sufficiently alien that we should fear it? The image really does become Mephistophelian (or Californian) enough that we ought to worry about the extent to which we have lost touch with reality.

The existence of a technology does not entail that it will be used. We all know people who refuse to use computers today, not because they can't, but simply because they feel more comfortable with the old technology of pen and paper. Surely we don't want to say that these individuals are controlled by pencil and paper! The decision to employ a certain means to an end requires thought, information, a determination of the nature and desirability of the end, assessment of the long- and short-term costs and benefits as well as constant updating of the data base. What if, in his or her declining years, our pen-and-pencil advocate changes his or her mind and opts for the computer, having decided that time is running out and there are too many things to finish by hand? Do we really want to say the machine won out over humankind? Surely not; an individual initiated the process that led to the machine, so why not include the individual in that process?

We are at the point where, in closing, we might ask: Why are we so quick to point to the machines and wag our finger? Well, the long and short of it is that those who fear reified technology really fear other individuals. It is not the machine that is frightening, it is what some individuals will do with the machine; or, given the machine, what we fail to do by way of assessment and planning. It may be only a slogan, but there is a ring of truth to: "guns don't kill, people do." There is no problem about the autonomy of technology. The problem is with individuals. The tools by themselves do nothing. That is the only significant sense of autonomy you can find for technology.

NOTES

1. John McDermott, "Technology: The Opiate of the Intellectuals," *The New York Review of Books*, 31 July 1969, 25–34.

2. Cf. Langdon Winner, *Autonomous Technology* (Cambridge: MIT Press, 1979).

3. Joseph C. Pitt, "The Epistemological Engine," *Philosophica* 32 (1983): 77–95.

4. S. Drake, *Galileo at Work* (Chicago: University of Chicago Press, 1978).

5. Ibid., p. 139.

6. Ibid., p. 141.

7. Ibid., p. 140.

8. I have worked some on the topic as have McMullin, Shea, Mittelstrass, and Wallace. Cf. Joseph C. Pitt, "Galileo: Causation and the Use of Geometry," in *New Perspectives on Galileo*, ed. Robert E. Butts and Joseph C. Pitt (Dordrecht: D. Reidel, 1978), pp. 181–95; "Galileo et La Spiegazione Razionale," in *I Modi Del Progresso*, ed. Marcello Pera and Joseph C. Pitt (Milan: Il Saggiatore, 1985), pp. 119–36.

9. The inventor, Hans Lipperhey, was a lens grinder; the invention was apparently the result of simply fooling around with a couple of lens, the basic properties of which were known through Lipperhey's daily experience.

The Labor-Saving Device: Evidence of Responsibility?

EDMUND F. BYRNE

It is unquestionably people who decide to develop and introduce new technologies. But technology advances so rapidly now that it is no longer apparent to the uninvolved observer that human beings are actually still in control of it. Technology, it is said, is on a "runaway" course; it is becoming "autonomous" of human agency. Taken literally, this characterization of technological change leaves no theoretical room for moral if even for legal responsibility. In the absence of any more subtle qualifications, it amounts to an instantiation of the sort of hard determinism that rules out the very possibility of free will.[1]

The case for technological determinism can, to be sure, be argued with a brief full of indications that technology is "out of control." Conceded. Moreover, indications that human beings are still in control can be denied theoretical import by asserting the "it's-only-a-matter-of-time" principle. Also conceded. Yet it remains the case that *some* human beings are clearly responsible agents of technological change. By this I mean that some human beings not only consciously seek the development of certain technologies but tolerate or even intend fully anticipated consequences of the introduction of that technology. I believe that this claim can be generalized so as to apply in some degree to any technological change. But here I shall build my case only on the subset of technologies developed and introduced as "labor-saving devices."

Assumed for purposes of this discussion is that workers are in fact displaced and are thereby harmed by the introduction of labor-saving devices. To be established is the claim that in each instance some human beings as agents consciously intend, at least indirectly, the harm that does result from the introduction of so characterized technologies. In support of this claim I propose to discredit each of four counterarguments:

1. Only good is intended by the introduction of a labor-saving device, as witness the support of workers as well as management.
2. If workers are opposed to doing away with work, then it is demonstrably false that some people intend the unemployment consequence of labor-saving devices. For management surely has a long record of encouraging and stimulating hard work.
3. Even if some people do intend to reduce the work force by intro-ducing a labor-saving device, they are not acting freely but only as they are compelled to do in response to the competition.
4. Even if some human beings do intend the unemployment con-sequences of labor-saving devices, this does not prove they are acting against the interests of workers. For the long-term result of this process will be a state of affairs in which people will no longer need to work.

Even if I am able to neutralize these objections, I will not, of course, have established any universal claim with regard to responsi-bility for technology. But the evidence I propose to introduce with regard to labor-saving devices might at least serve as an invitation to the opposition to assume the burden of proof.

1. LABOR-SAVING AS A EUPHEMISM

Only good is intended by the introduction of a labor-saving device, as witness the support of workers as well as management.

Response: Appeal to worker support is misleading, because the notion of a labor-saving device is (a) ambiguous and (b) euphemistic.
Saving labor does not have the same meaning in everybody's vocabulary. To management saving labor means reducing produc-tion costs by lowering wages and/or benefits, or by shrinking work-ing hours and/or the work force itself, by automation or whatever. Working people, on the other hand (at least those who are not poli-tically astute), typically associate saving labor not with any reduc-tion in employment but with a reduction in the amount of *hard,*

debilitating work that they must do to earn a living and/or carry out their domestic chores.

There is, then, both agreement and disagreement between management and workers with regard to the import of "labor saving." All agree that what should be saved (i.e., reduced or eliminated) is tedious debilitating work. Disagreement, and thus misunderstanding, has to do with the relevance of such labor saving to jobs. Workers generally favor reducing or eliminating not wages or jobs but only the demeaning aspects of work. Management, however, typically favors reducing or eliminating both the demeaning aspects of jobs and the holders of those jobs.

Evidence for this latter claim includes at least the following three observations. First, elimination of unpleasant aspects of work is typically cited as a reason for adopting new technology.

Second, in the absence of new technology, management typically appeals to reasons for doing the work anyway. But to the extent that the work in question is unpleasant, the reasons cited need to overcome what is probably a natural aversion on the part of the workers. So most attempts to "sell" labor to workers depend upon some special gimmick.

Perhaps the most daring gimmick of all, utilized by Voltaire and by some German philosophers in the nineteenth century, is to declare flat out that labor is wholesome, meaningful, and fulfilling. Few defenders of labor have been quite so daring. Most have taken the softer approach of emphasizing the instrumental value of work as a means to some end. Work might be praised as the way to a better world, for example, in the Marxist perspective. Work (associated particularly with labor) might also be recognized as painful but valuable for that very reason because of some indirect benefit that arises out of the pain. This approach has been widely utilized in the West, especially in the form of a "work ethic" that encourages people to work hard by appealing to their interest in being saved—not in this life, but in a promised life to come.

Third, even jobs that are not inherently or even predominantly unpleasant are continaully being eliminated when and as new technology becomes available. This in itself is a strong indication that the notion of a labor-saving device, as understood by management, is a euphemism. This, however, will become clearer from considerations below of the antilabor bias of management.

2. MANAGERIAL SUPPORT OF HARD WORK: THE INTERIM RULE

If workers are opposed to doing away with work, then it is demonstrably false that some people intend the unemployment consequences of labor-saving devices. For management surely has a long record of encouraging and stimulating hard work.

Response: The appeal to hard work is only an interim device that, at least since the Industrial Revolution, is relied upon while awaiting a technological fix.

Work has long been a basic feature of social planning all across the ideological spectrum. For in a world in which at least some human beings need to work in order that their kind can survive and occasionally even prosper, it is difficult for a social planner not to take work into account. But in taking work into account, two different but interconnected questions need to be addressed: (1) what work ought to be done; and (2) who (or what) ought to do it. The answer to the first question depends on goal selection—for example, with regard to desired level of need satisfaction and/or commitment to progress. The answer to the second at least presupposes a theory of distributive justice in society. More egalitarian ideals have been espoused in what I call the communitarian tradition.[2] But proponents of labor-saving devices are more likely to be associated with the authoritarian tradition.

Linked together as complementary themes of the authoritarian tradition are (1) the claims of an elite to exemption from some or all work, and (2) some rationale to justify imposing the burdens of work on a subservient class or classes of people. The principal reason given for requiring work of others has always been that they will not otherwise qualify for social benefits. This reason has, in turn, been mystified with reminders that people are being punished for something or other and/or are on probation for a less demanding life ahead. Typical reasons given for excusing oneself from work that one requries of others is that one has more (socially) important things to do or, simply, that one is "the boss."

Authoritarian strategy with regard to the work force has long been characterized by an interim rule and an ultimate rule. The interim rule is: do what you must to get productivity out of your workers. The ultimate rule is: whenever possible, replace people with machines. The history of the interim rule reveals the origins of labor saving in antiworker bias. The history of the replacement rule shows directly the culmination of antiworker bias in worker-displacing technology and indirectly the conceptual inadequacy of the technological fix.   .

German sociologist Max Weber identified the making of money

as the single-minded, joyless, and ultimately irrational goal of the work ethic.[3] And as he described it, this ethic could be interiorized only by the actual or would-be entrepreneur. There is, however, a corollary to the work ethic to the effect that poverty is not circumstantial but is a direct result of its victim's failure or refusal to work. In this latter form, the work ethic has been applied even to the unpropertied wage laborer as an incentive to greater productivity, by means as diverse as piece work and workhouses. As thus "applied," the work ethic has served as an interim strategy of the authoritarian philosophy of work. It is generally associated with the Protestant reformer John Calvin and the Calvinist Puritans. But it did not originate with Calvin; and its appeal transcends ideological boundaries.

Scholars following the lead of Marx and Weber have generally assumed that capitalism developed in the sixteenth century; that before that time society had been made up mainly of peasants for whom the principal economic unit was the extended family or, at most, the village; and that the destruction of the latter was more or less a sine qua non for the occurrence of the former. It is now recognized that socioeconomic realities were notably more complex than this, and that, in particular, the task of developing theological justifications of trade and commerce was already well underway in the thirteenth century. Up until that time, however, literate Christians in Europe tended to be as elitist and authoritarian on the subject of work as were their cultural forebears.[4]

Saint Benedict (d. 543) is retroactively credited with having made work and prayer (*ora et labora*) the dual objectives of the monastic order that he founded in Italy; and on this basis he is said to have restored to work the dignity that ancient civilizations had denied to it. This interpretation is, however, too simplistic. If anything, Benedict stands in history as a dutiful perpetuator of the values of authoritarian rule. He has no respect for "sarabites," because they are monks who consider whatever they think good or choice to be holy and whatever they do not wish to be unlawful. The life style of "gyratory" monks, who are "always wandering and never stationary" (who, in a word, imitate in this respect Jesus and his apostles) is, in a word, "wretched." Only cenobites, who live together in total and unquestioning obedience to their abbot, are truly worthy monks. In his renowned and influential "Rule" he spells out just how the hours of the day are to be divided up, in the different seasons, between manual labor and sacred reading. The former is never to be so "violent" as to drive away the more feeble or delicate brothers; but, he also notes, "they are truly monks if they live by the labours of their hands; as did also our fathers and the apostles."[5]

Whatever Benedict's intentions, in the course of time this early commitment to manual labor was reduced to mere symbolism as the monks left hard work to serfs and wage earners and engaged themselves only in more honorific and less tiring endeavors such as baking, gardening, and brewing.[6]

The holy elitism that characterized "reformed" Benedictinism is also found among the (clerical) intellectuals of the high Middle Ages who depended on institutionalized thinking to earn their living. Thomas Aquinas, a thirteenth-century Italian monk with aristocratic origins, repeated Benedict's endorsement of both contemplation and actions, but was willing to separate one from the other on the basis of one's position in society. He agreed with Aristotle that contemplation is the highest human endeavor, to which the chosen few might devote all their attention. And he also agreed with St. Paul's admonition, "If any man would not work, neither should he eat"— but only to the extent that such work is necessary. Where it was "necessary" was among the common people, who accordingly had a right as well as a duty to work. They were not to question or attempt to rise beyond their "natural" station in life; but in turn they were entitled to a "just price" (*justum pretium*) for their labor—just enough to provide a bare livelihood for one's self and one's family.[7]

This essentially authoritarian view of appropriate compensation for work did accommodate a paternalistic exception for the disabled. Anyone who was truly unable to work was recognized as having for that very reason a right to be cared for. But it was no more obvious in the Middle Ages than it is today just who is sufficiently disabled to merit care without working. At the time of Thomas Aquinas there was, to speak anachronistically, a sizable surplus pool of labor. And as a result begging was an accepted means of gaining one's livelihood—one to which the formerly wealthy St. Francis of Assisi gave spiritual dignity by requiring his monks to rely upon it for their daily bread.

This comparatively idyllic policy with regard to the work obligation came to an end in the first half of the fourteenth century, for reasons that are still not clearly understood. But just as commerce was beginning to expand in various directions, western Europe was decimated first by famine and then by the Black Death, which in 1347 swept across the continent from Constantinople and by 1349 had eliminated over a third of the population of England. The resulting sharp reduction in the labor supply led to the Statute of Laborers (1349), which accommodated English landowners' need for agricultural workers by forbidding the able-bodied to beg, travel, or demand more than customary wages and requiring them to labor for their livelihood. Subject to the penalty of imprisonment, unskilled people who survived the plague were thus circumscribed:

> That every man and woman of our realm of England, of what condition he be, free or bond, able in body, and within the age of three-score years, not living in merchandize, nor exercising any craft, nor having of his own whereof he may live, nor his own land, about whose tillage he may occupy himself, and not serving another, if he in convenient services, his estate considered, be required to serve, he shall be bounden to serve him which so shall him require; and take only the wages, livery, meed, or salary, which were accustomed to be given in the places where he oweth to serve, the twentieth year of our reign of England, or five or six other common years next before.[8]

Or, as one modern commentator puts it, "The King and his lards saw begging, movement and vagrancy, and the labor shortage as essentially the same problem, to be dealt with in one law."[9] But the fabric of feudalism was coming undone, and the serfs' quest for freedom generated more and meaner laws. Persons without a letter authorizing travel were to be put in the stocks (12th Richard II, 1388). Yet after a century of thus battling against "idleness, mother and roof of all vices," it was noted at the time of King Henry VIII that the number of vagabonds and beggars had actually increased. So provision was made for those truly in need (statute of 1531); but able-bodied loafers were subject first to public whipping in the nude, then to whipping plus loss of part of one's right ear, then if still not willing to "put himself to labor like as a true man oweth to do," to "pains and execution of death" (statute of 1536). During this same period of time most European countries empowered a new official, known in England as overseer of the poor, to put poor people to work and to imprison those who refused or performed unsatisfactorily (a somewhat primitive approach to vocational training). Meanwhile increasingly severe laws were being enacted to outlaw begging.[10]

It is in this context that Protestant reformers put forward their views about work. Martin Luther (1483–1546) still drew upon the just-price theory to justify telling people to work at the trade or profession into which they were born. But he attributed equal value to any kind of work, active or contemplative, and stressed the religious dignity of one's work as a vocation or calling. Thus, in his little book about vagabonds, *Liber Vagatorum*, he linked the Reformation to the growing movement against beggars by endorsing almsgiving only to duly certified indigents. The Lutheran Eberlein proposed in his utopian *Wolfaria* (1521) to abolish serfdom, execute all mendicant friars, strictly regulate all trades and professions to avoid production of luxuries, and set everyone, including the nobility, to work at the only really honest occupation: agriculture. Johann Andreae, a Lutheran priest, dreamed of a society called *Christianapolis* in which perfect officials would not tolerate begging and would give material assistance to the poor only after careful examination of their needs.[11]

John Calvin (1509–64) pushed the significance of one's work even farther by tying it in some inscrutable way to one's eternal salvation. So casual work is for this purpose inadequate, and dislike of work raises serious doubts about one's being among the elect. Although committed to a rigorous doctrine of divine predestination according to which human choice is irrelevant to the final outcome, Calvin insisted that the faith by which one is saved is expressed in and through methodical, disciplined, rational, uniform, and hence specialized work. Puritanism, which was an offshoot of Calvin's teachings, drew the logical conclusion that wealth seeking is a fine way to assure one's salvation; and in this way, according to Max Weber, Calvin's austere theology provided the ideological underpinning for capitalism. According to another interpretation, however, what Calvin provided was a religious justification for the capitalist's hard-nosed approach to discipline on the assembly line.[12]

The latter interpretation is certainly borne out by the technocratic moralizing of Scottish engineer Andrew Ure. Ure, the Calvinist ideologue of the Industrial Revolution, explained the need for hard work by associating its pain with that of the crucifixion of Christ.[13] But Calvinism was not alone in its endorsement of work either during the mass-production or the mechanization phase of the Industrial Revolution. Robert Burton, a British don with no known religious preference, dreamed in his *Anatomy of Melancholy* (1621) of a society in which there would be no "beggar, rogues, vagabonds, or idle persons at all, that cannot give an account of their lives how they maintain themselves" and in which all able-bodied poor would be "enforced to work."[14] Two centuries later, Joseph Proudhon developed an anarchist glorification of labor around the idea that the value of work is directly proportional to how hard it is.[15] Karl Marx, by comparison, was not nearly so enamored with work, as will be noted. But Andre Gorz, a contemporary Marxist, has said that "after the communist revolution we will work more, not less."[16]

Of course, these and other encomia of work, whatever their ideological roots, share a common flaw: even if a society depends on the work of some, that work will not be held in esteem, either by the workers or by their beneficiaries, if something else, notably ownership, is considered more honorific.[17] But the point here is simply that a work ethic serves the purpose of encouraging productivity in the absence of suitable machines. This is evident, for example, from the various work-force development proposals put forward in England during the sixteenth and seventeenth centuries. Faced with numerous and potentially revolutionary poor people, literate members of the leisure class came up with all sorts of ideas about how best to put the poor to work without upsetting the rich. Typically conservative ideas for social experiments of every sort, on both the community and the national scale, aimed at doing away with vagabonds and beggars, turning them into totally responsive instruments that, properly organized, would turn a profit for the enterprising rich.[18] This organization of "manu-facture," which culminated in Ambrose Crowley's authoritarian Law Books for his ironworks, won the praise of eighteenth-century landowner Adam Smith, who has immortalized "mass production" of pins. But even as this "assemblying" of workers became common, the specter of "labor-saving devices" loomed on the horizon.

Gabriel Plattes argued that such labor-saving devices should not be adopted until a labor shortage has first developed. But John Bellers, a wealthy Quaker professionally concerned with poor relief, took the view that prohibiting a labor-saving device by law is like requiring a laborer to work with one hand tied behind his back. Despite some legal approaches of the sort favored by Plattes, it was Bellers's openness to technology that served as a model for the entrepreneurs to come.[19]

3. MANAGERIAL RESPONSIBILITY FOR "LABOR-SAVING": THE ULTIMATE RULE

Even if some people do intend to reduce the work force by introducing a labor-saving device, they are not acting freely but only as they are compelled to do in response to the competition.

Response: (a) Labor-saving devices are not necessarily adopted in response to competition; and (b) even if concern about competition is a consideration, a labor-saving device may not be a suitable response.

Already in the Middle Ages people took delight in various devices that could do something ordinarily done by humans. But in the absence of motivators like profit and progress, such devices were perceived merely as objects of wonder. As capitalism gradually transformed social goals, people learned to think of such devices as sources not just of wonder but of lower-cost productivity by means of which they might gain an advantage over their competitors. As we shall see, this is not necessarily so. Besides, labor saving has not been espoused solely to beat the competition. It has also been motivated by a desire to separate workers from their jobs—either directly, as a result of a deliberate antiworker bias, or indirectly, as a result of a protechnology bias the consequences of which for workers are not well anticipated. The indirect bias will be considered in connection with the fourth counterargument. Here only the direct bias will be considered, before addressing the relevance of competition.

The Direct Antiworker Bias

Almost from the onset of the Industrial Revolution, with its characteristic centralization of workers in capitalist-controlled plants, forces were set in motion that made the idea of production without payrolls attractive to the entrepreneurial class. Skilled workers who had previously enjoyed relative autonomy in their work life chafed at the impoverishing terms offered to them by the factory owners. The response of the owners, as often as not, was to look for a technological substitute for such intractable employees.

Thus, for example, did a certain Mr. Roberts respond to British textile entrepreneurs by developing a spinning automaton known as "the Iron Man" to displace high-wage skilled spinners. More generally, according to Andrew Ure,

> wherever a process requires peculiar dexterity and steadiness of hand, it is withdrawn as soon as possible from the *cunning* workman, who is prone to irregularities of many kinds, and it is placed in charge of a peculiar mechanism, so self-regulating, that a child may superintend it.

In this way, he notes, "when capital enlists science in her service, the refractory kind of labour will always be taught docility."[20]

A decision to replace skilled workers with some labor-saving device is seldom if ever going to be made, of course, just out of a desire for docility of the sort articulated by Ure. In the first place, science is not always prepared to provide industry with a quick technological fix, nor is labor saving the only motive for introducing new technology. A substantial part of good management strategy in this regard is to determine what is most advantageous in light of all known variables. Having determined this insofar as possible, one might in a given situation choose not to introduce an available device—until, that is, there is a significant change in one of the variables.

Standard dogma in this regard among neoclassical economists was to the effect that a rise in the cost of labor will precede a decision to mechanize. Introduction of "the Iron Man" into the textile industry is an example. Others are provided by Karl Marx. He notes how manufacturers in England turned to mechanization only after the Factory Laws limited child labor to four- to six-hour shifts and children's parents refused to sell their "half-timers" for less than full-timers. He also points to the practice of producing machines in one country to be used in another country where high wages motivate such substitution, and indicates that this very practice so expands the labor pool in the country where the machines are introduced that other industries there are spared the need to mechanize. For, says Marx, the capitalist's "profit comes . . . not from a diminution of the labour employed but of the labor paid for."[21] Thus, even where displacement of workers by machines is prohibitively expensive if not still technically unfeasible, the very threat of doing so might be employed to dissuade workers from demanding higher compensation.[22]

The need to make a profit is unquestionably an important reason for a company to look for ways to cut costs; but labor saving is only one of the ways to cut costs, other possible ways being through cheaper raw materials or cheaper service on corporate debt. Besides, there are factors other than profit or revenue maximization that management might consider to be of overriding importance. For example, it is reported that a landowner in India will resist profit maximization of his agricultural business if by increasing his tenants' share he risks having his tenants pay off their debts to him and get out from under his control.[23]

That control over the work force can even take precedence over profit maximization is central to the Marxist analysis of mechanization under capitalism. In Harry Braverman's view, for example, control of the work process is the overarching reason for mechanization. The "deskilling" of production that mechanization had effected is, according to Braverman, deliberate but not inevitable. In keeping with the Marxist class analysis of this process, he too speaks of capitalism striving for "domination of dead labor [machinery] over living labor [workers]," thereby attributing an antilabor animus to the business decisions to mechanize and automate.[24] The evidence he is able to muster for this claim is largely circumstantial and anecdotal—for example, the fact that supervisor-operated numerical control (N/C) became the automation of choice rather than the equally efficient record-playback system (R/P) that leaves programming in the control of skilled machinists.[25] Historian David Noble has since elaborated upon this example in impressive fashion, but without finding a "smoking gun."[26] Other Braverman inspired studies of the labor process also lend support to the claim that skilled industrial workers are being systematically displaced by machinery.[27]

Whether and in what numbers workers affected by deskilling ever rejoin the work force is a subject of much debate. What seems clearly beyond debate is that the immediate purpose of a labor-saving device is, as its very name indicates, to save labor—this is, to save a company some of the costs associated with paying labor. This means that the process of introducing such a device is inherently even if not intentionally hostile to anyone whose labor is thereby going to be "saved." Marx and many who have followed him would go on to argue that there is a definite and deliberate antilabor bias on the part of management that tilts cost-cutting decisions wherever possible in the direction of cutting payrolls.

A generalized claim to this effect is difficult if not impossible to prove. (Braverman says this is because the complexity of reasons why there has been a "transformation of the labor process" does not lend itself to a "unitary answer.")[28] But many on the side of management, especially industrial design engineers, have displayed just this sort of antilabor bias. Ure, for example, put it thus: "It is, in fact, the constant aim and tendency of every improvement in machinery to supersede human labor altogether, or to diminish its cost, by substituting the industry of women and children for that of men; or that of ordinary labourers for trained artisans."[29]

The century and a half since Ure has been characterized by a variety of strategies to hold down the cost of labor in manufacturing, including the total regimentation of company towns and Frederick Taylor's preference for "stupid" workers who can be counted on to be docile, advancing to various managerial theories of the twentieth century and culminating in an all-out effort to design workers right out of production processes entirely.[30]

As this trend clearly indicates, technological unemployment is inevitable, to a degree still subject to debate. It is inevitable, however, not by virtue of any law of nature or because there are no alternatives with different consequences but because robots and other microelectronic devices are already perceived as cost effective in the long run and hence a necessary condition for staying competitive in the industries affected.[31] Thus is being carried out Ure's "automatic plan," which he defined as follows: "Skilled labour gets progressively superseded, and will, eventually, be replaced by mere overlookers of machines."[32] And to this day engineers seem to believe that automation can be cost effective only by eliminating humans, because the greatest expense is incurred in trying to accommodate "man in the loop." Says Lawrence B. Evans, an MIT chemical engineer:

> The cost of complex electronic circuitry continues to decrease exponentially (by a factor of about 1/2 each year) due to large-scale integration (LSI) semiconductor technology. . . . The real cost of a system is in the hardware for communication between man and that system (displays, keys, typewriters) and this cost is a function of the way the system is packaged. Thus, automation functions and data processing become economic if they can be done blindly, without the need for human communication.[33]

Estimates vary as to just how much less expensive it may be to use robots in place of humans; but that there will be significant savings is widely assumed. As one writer puts it, a Japanese robot in automotive production can do at $4.50/hr. what a member of the United Automobile Workers does for $18.10/hr. (wages and fringes).[34] An estimate of this sort is typically based on a comparison between costs incurred from labor and costs of procuring and maintaining a robot. Robot providers claim that robot costs will be recouped within a three-year payback period from savings in labor alone. Of course, assumptions with regard to the cost of money, the cost of installation, and the cost of power and maintenance of a robot need to be adjusted up to take inflation into account. But the initial cost of producing a robot may well drop from, say, $50,000 in 1980 to just $10,000 in 1990. So recent estimates are probably at least in the correct order of magnitude.

Technology and Competition

The foregoing are clearly indications of free choice in the decision to adopt a labor-saving device. But, it may be objected, they are not persuasive because they ignore the background conditions that mandate such as decision in the first place. These conditions, it is argued, involved competition among firms, each of which seeks to control or at least to constrain others. It does not follow from this, however, that a labor-saving device is the appropriate response.

The labor-saving device has acquired a reputation for being able to deliver a technological fix to any business or industry that has what is considered to be an excessive payroll. This reputation, in turn, enhances significantly the success rate of suppliers who want to sell labor-saving devices to corporate buyers. What, if anything, the latter or anyone else stands to gain from any such purchase is, however, by no means obvious.

That society may wind up worse off because of corporate commitments to automation may be seen either on the level of the affected individuals or on the level of society taken collectively. It is beyond dispute that individual workers are harmed by automation and that society is seldom able to undo all the harm. And as for society, the automation decisions of corporations are not made with a view to benefiting society as a whole. With a view rather to making money for their investors, companies being tempted by automation take only their own ("internal") costs into account, not the overall ("external") costs, direct and indirect, that spill over onto society in the wake of major technological change. The overall costs, most of which the comparatively defenseless members of society are usually called on to bear, can be traced for the most part to the loss of jobs in a society that distributes benefits on the basis of one's employment.

What is especially deserving of our attention is the question of whether, or under what circumstances, technology in the guise of a labor-saving device is likely to benefit its corporate buyer. One might suppose in this regard that any corporation that could consistently answer these questions would know how to allocate research and development money for the purpose. But the matter is not that simple, in part because of uncertainty about the performance of the device and in part because of uncertainty about the performance of competitors who are also in a position to adopt a functionally similar device.

Why there must be uncertainty about the performance of a device is abundantly illustrated in the history of technology, for example, the development of machines to do calculations, from the counting board to the computer.[35] Inventors such as Leibniz, Pascal, and Babbage produced mechanical devices that could do calculations accurately, but not rapidly enough to make it economically attractive to substitute them for human clerks and accountants. Electricity introduced a new factor into the equation, but further development had to await improved designs as well as a situation of cost-discounting urgency. The Allies in World War II needed to expedite the development of trajectory charts for new types of shells, which were being introduced into the arsenal faster than human calculators could track them. The needs of the war effort outweighed expenses as the ingenious ideas of Vannevar Bush and others were applied to produce the first workable computer. That it was able to perform its tasks as expected is, however, irrelevant to considerations of economy. There simply was no upper limit of expense for such research under wartime conditions. This situation was not without precedent, of course, and it continues to this day—not with a view to winning a war, necessarily, only with a view to "defense."

Economically unjustified introduction of labor-saving devices does not occur with such indifference to cost in the private sector. But this does not preclude a corporate commitment to high-risk adventure, for reasons having to do with everything from the corporate image to the corporate will to dominate an industry. The former is often a shortcut to bankruptcy, the latter, a strategy best left to oligopolists. Thus, for example, might narrow-gauge cost considerations about the introduction of robots give way on occasion to a desire for product-quality improvement—for example, in production of Chrysler's K-car at the Newark, Delaware, plant and of General Motors' Fleetwood in Detroit, where $8.5 million of robots save only $120,000/yr. In such instances, a more affluent market is targeted, and cost is expected to be recouped through sales.[36]

It is in this context that one needs to consider neoclassical economic theories about motivation for technical change.[37] On the assumption that the entrepreneur rationally selects that mix of labor and capital that will maximize net revenue or profit, the orthodox position before World War II was that labor-saving innovations come about as a direct result of high-priced labor (and similarly with regard to the price of capital). The logic in that position is faulty, because total costs of production can be brought down by reducing the cost of either labor or capital. So there would be no special reason to concentrate on reducing the cost of labor unless one has either easier access to labor-saving knowledge or, as in the case of Ure, an antilabor(er) bias.

The orthodox position is even less persuasive if confronted with game theory considerations about the disparity between collective and individual benefits to be derived from a labor-saving innovation. If it be assumed that no entrepreneur knows what any other is going to do, each is caught in a prisoners' dilemma over how to avoid being the only one to bear the cost of introducing, or of not introducing, the labor-saving innovation. If, in addition, it be assumed that a search for cost-reducing innovations will be the same whether one focuses on saving labor or on saving capital, then in the absence of some special reason to favor labor-saving, such as a history of rising labor costs, the choice of direction might depend on nothing more than a flip of a coin.

Suppose, finally, that, an entrepreneur knows that one or more competing entrepreneurs might invest in a labor-saving device.

Alteris paribus, it is still not necessarily the case that the entrepreneur armed with this information should invest in the same labor-saving device. For, if one's competitors do in fact "save labor" with their new device, the cost of labor in the industry will fall, and our noninnovating entrepreneur will be able to undercut the competition by virtue of both having avoided the cost of innovation and benefiting from the availability of cheaper labor. Because, however, the other entrepreneurs can be presumed to be making the same assessment, the entrepreneur must also contend with the possibility that none will innovate, in which case the cost of labor will continue to rise and the advantages or innovating unilaterally will increase. In short, introducing the labor-saving device would be the rational profit-maximization course of action only for the entrepreneur who does so unilaterally, not for those who "follow the crowd." (What better counterargument to the old saw: Be not the first by whom the new is tried nor yet the last to cast the old aside?)

The resulting quandary about whether to invest in the labor-saving device has been described as a game of chicken. It is such, however, only if (1) the innovation in question merely substitutes for rather than improves upon the product and/or its production and (2) no competitor does in fact make a move to introduce the innovation. Practice with regard to introducing innovations into traditionally labor-intensive industries in the 1980s satisfies, or at least is perceived as satisfying, both of these provisos. So in the real world that theoretical quests for perfectionality have so far failed to model, the game being played is not chicken, but corporate survival; and the only thing certain is that the work force will not be the winner.

4. THE EUDAEMONIC RATIONALE

Even if some human beings do intend the unemployment consequences of labor-saving devices, this does not prove that they are acting against the interests of workers. For the long-term result of this process will be a state of affairs in which people will no longer need to work.

Response: The end envisioned is not easily extrapolated from available data; and, this notwithstanding, felicitous ends do not justify unjust means.

Working-class endorsement of the saving of labor is not a spontaneous tropism on the part of those who live too close to labor. It has been instilled over the centuries by promises of a work-free utopia to come. Current versions of these hoary promises go beyond their classic predecessors by noting a need to reeducate people accustomed to labor for a life of leisure. There are basically just two versions of the utopian scenario. One recognizes a painful transition as we pass through the tunnel on our way to the light. The other, notably more pollyanish, skips over interim angst to concentrate on the latter days.

The rationale here runs more or less as follows. For the first time in history technology has brought us to a point at which there are not and will never again be enough jobs to go around. So, whatever may have been society's need for workers in the past, that need is with us no longer. We have reached a point that has all along been proclaimed as a goal of technological innovation, namely, to eliminate the need for "labor" (meaning paid workers) by utilizing machines instead. Only so long as technology could not deliver on this promise, the obsolescence argument would continue, was there any need to instill in *humans* a sense of duty with regard to work. Now this need is passing from the scene, so society must transform its values and its objectives accordingly.

The pollyanish proposal, presented as a prediction, is that society must begin to focus not on work, which is no longer an appropriate goal, but on leisure, which we shall have in the future whether we are ready for it or not. In other words, work is less valuable now than it once was because it is less important to the satisfaction of human needs than it used to be. Robert Theobald, for example, envisions the emerging situation in this way:

> [In] the new society we are entering. . . many people will tend to work intensively for a period of time and then need to re-create and re-educate themselves. We shall extend rapidly the concept of sabbaticals in terms of the number of people involved, the number of occupations for which they are considered relevent, and the length of time for which people can free themselves up from responsibilities. Societies will be able to make free time available because of the impacts of computers and robots, which will limit the amount of human energy needed for industrial-era jobs.[38]

None of this, by the way, would have saddened Marx and his followers, provided only that control of the means of production be not in the hands of the capitalists but of the proletariat as represented by the State. Marx in his youth envisioned a future in which machines would be doing the work and the only question would be who was going to reap the benefits. He did not, to be sure, think of the road to such a state of affairs as one to be easily traversed. Respectfully rejecting the escapist proposals of the utopian socialists, both Marx and Engels sought to confront the new industrial reality head-on to salvage a future for the vast majority of human beings who make up the working class. On this view, labor, however valuable in the capitalist setting, has no intrinsic value. Machines are welcome as a means to the eventual liberation of human beings from dehumanizing drudgery. Lenin in the interim welcomed even Taylorism as an appropriate device for increasing productivity.[39] Workers are alienated from their work in the typical factory system, but this is due not to rationalization of work but to capitalist ownership. Scientific socialism promises the surmounting of alienation, first by assuring workers that collectively they own the means of production, and in time perhaps by freeing them of responsibility for production and handing this over to machines. Marxism, then, does not romanticize work but rather socializes the work ethic for the duration of our dependence on human labor for productivity.

In this regard, Harry Braverman acknowledges that no less deskilling mechanization has taken place in Communist countries; but he excuses deskilling there because it is merely imitative of what capitalists did and expresses the hope that in these countries the dominance of machines over people is only transitional.[40] An even more utopian expression of hope in this regard is that of another neo-Marxist, Herbert Marcuse, who welcomes automation in spite of the short-range concerns of workers. These concerns, says Marcuse, are legitimate in the absence of "compensating employment." But, he insists, over the midrange of time such opposition to technical progress prevents "more efficient utilization of capital," "hampers intensified efforts to raise the productivity of labor," and leads to economic crisis and exacerbation of class conflicts. That is bad enough. What is worse is that opposition to automation stands in the way of eventual attainment of a liberating utopia based on technology: "Complete automation in the realm of necessity would open the dimension of free time as the one in which man's private *and* societal existence would constitute itself. This would be the historical transcendence toward a new civilization."[41]

Neo-Marxist Andre Gorz is also convinced that technology has prepared "paths to paradise." He is careful to point out, however, that these paths will be traversed toward "liberation from work" only "within a social environment which does not yet exist (at least one generally)."[42]

Whether the ultimate outcome of all of this will be anything like the downfall of capitalism that Marx predicted is a matter for specialists to debate. And as they do so they will want to bear in mind that the assumptions of classical economics, against the background of which Marx developed his alternative theory, are inadequate to represent the complexities of today's transnational marketplace of competing forces. What is important in the present context, somewhat more simply, is the impact that the delaboring of productivity will have on people regardless of the sociopolitical framework within which their onetime workplaces happen to have been located. On this issue divergent ideologies are of only secondary importance. More basic is the cross-cultural heritage of the human race with regard to the value and necessity of work, which finds its way into the views of writers of every persuasion.

Take, for example, Herbert Marcuse's guarded optimism about the benefits that labor-saving technology will bestow upon the working class. That Marcuse places so much trust in the eventual blessings of technology for the working class reflects his interest in Marx's earlier writings.[43] But it does not reflect his interest in the writings of Freud. Although Freud seldom addressed the subject of work directly, one footnote in his *Civilization and Its Discontents* states explicity what elsewhere is only implicit. Work, he says here, is the best means of tying an individual to the community and, if it is work at a profession, is an excellent instrument of sublimation. But, he regrets, "as a path to happiness, work is not highly prized by men. They do not strive after it as they do after other possibilities of satisfaction. The great majority of people only work under the stress of necessity, and this natural human aversion to work raises most difficult social problems."[44]

Thus is suggested the view that work, even if eventually unnecessary for economic productivity, is nonetheless an important vehicle of human creativity. This view underlies the concerns of Eric Fromm about the possible demise of work.[45] And a century earlier it underlay William Morris's aesthetic of work and Proudhon's glorification of work as having intrinsic dignity.[46] It was important to William Wordsworth and to John Ruskin as they watched cottage industries giving way to dehumanizing division of labor in factories. And in our own day it is finding expression in the works of novelist and poet Marge Piercy.

Emphasis on the human need for work is a key feature of E. F. Schumacher's insistence that we move towards "appropriate technology." As he once expressed his ideological assumption, at least for the poor man, "the chance to work is the greatest of all needs, and even poorly paid and relatively unproductive work is better than idleness."[47] Schumacher himself tried to incorporate this pro-work view into a kind of Buddhist economics that stresses the importance of work to the individual and to the community. But it would not be difficult for the authoritarian tradition to co-opt such

humane theorizing for ends quite unrelated to anything that Schumacher wanted for the world. It was, after all, just this sort of co-
opting that lay at the foundations of the modern factory system. The
disciplined religious community that Saint Benedict developed in
the Sixth Century became a thousand years later a model for organized "manufacture" that was well in place long before the appearance of mechanized assembly.

These worries aside, there is still good reason to support the claim
that work, even laborious work, is a valuable instrument of human
fulfillment. This admittedly has not been proved to be true universally and without qualification. And, like claims with regard to our
reliance on gravity before humans experienced weightlessness, its
truth may turn out to be of somewhat limited applicability. But just
to the extent that it still obtains in the world as we know it, the
delaboring of production should not be endorsed without serious
qualification. In particular, it should not be endorsed so long as
work remains the principal means of support for oneself and for
one's dependents while at the same time welfare rights are viewed
not as the fruit of human progress but only as a recipe for sloth.[48]

This sort of cultural "catch-22" is not inherent in our genes. It is,
if you will, the result of holding on to the interim rule even though
the ultimate rule is now fully operative. Of far greater significance,
however, is the fact that human beings—and, not infrequently,
identifiable human beings—are responsible for the perpetuation of
these rules. For this constitutes a prima facie case of responsibility
for the consequences of their application. A defense based on "business necessity" could no doubt be mounted. But, I submit, this
defense can and should be overcome by showing that there are
alternative social solutions.

NOTES

1. The hard determinism implicit in the view of Jacques Ellul is open to an
unusually subtle qualification. Only his "sociological" works requires a determinist
conclusion; his theological works offer an escape in Christian activism. See my
review in *Nature and System* 3 (September 1981): 184–88.

2. The communitarian tradition, especially as exemplified in utopian literature,
looks to reform and reconstitution of social organization as the proper remedy for
work-related inequities. See Frank E. Manuel and Fritzie P. Manuel, *Utopian
Thought in the Western World* (Cambridge: Belknap Press of Harvard University
Press, 1979).

3. Max Weber, *The Protestant Ethic and the Spirit of Capitalism* (1904–5),
trans. Talcott Parsons (New York: Scribners, 1958), p. 53.

4. See Jacques LeGoff, *Time, Work, & Culture in the Middle Ages*, trans. A.

Goldhammer (Chicago: University of Chicago Press Phoenix Edition, 1982); Alan Macfarlane, *The Origins of English Individualism* (New York: Cambridge University Press, 1978).

5. The Rule of St. Benedict, in *Select Historial Documents of the Middle Ages*, trans. and ed. E. F. Henderson (London: G. Bell and Sons, 1925), pp. 274–75.

6. LeGoff, *Time, Work, Culture in the Middle Ages*, p. 84; James W. Thompson, *The Economic and Social History of the Middle Ages (300–1300)* (New York: The Century Co., 1928), pp. 607–18.

7. Adriano Tilgher, *Work* (1st English edition of *Homo Faber*, 1930), trans. D. C. Fisher (New York: Arno Press, 1977), pp. 39–42.

8. 23 Edward III, Statute of Laborers, 1349, quoted by Karl de Schweinitz, *England's Road to Social Security, 1349 to 1947* (Philadelphia and London: University of Pennsylvania and Oxford University Presses, 1947), p. 6. See Henderson, *Select Historical Documents of the Middle Ages*, pp. 165–68.

9. Ibid.

10. Ibid., pp. 20–38.

11. J. C. Davis, *Utopia and the Ideal Society: A Study of English Utopian Writing 1516–1700* (Cambridge: Cambridge University Press, 1982), p. 72.

12. Tilgher, *Work*, pp. 59–60.

13. Andrew Ure, *The Philosophy of Manufacturers or an Exposition of the Scientific, Moral and Commercial Economy of the Factory System of Great Britain* (London: C. Knight, 1835), p. 423.

14. Davis, *Utopia and the Ideal Society*, p. 100.

15. Manuel and Manuel, *Utopian Thought in the Western World*, p. 771.

16. Andre Gorz, "The Tyranny of the Factory: Today and Tomorrow," in *The Division of Labor*, ed. Andre Gorz (Sussex, England: Harvester, 1976), p. 58. More recently, Gorz has bought into the dream of a technogenerated utopia. See below in connection with no. 42.

17. Cf. Thorstein Veblen, *The Theory of the Leisure Class* (1899) (New York: Modern Library, 1934), pp. 92–95, 231.

18. Davis, *Utopia and the Ideal Society*, chap. 11, pp. 299–367.

19. Ibid., pp. 318, 345–46.

20. Andrew Ure, *Philosophy of Manufacturers* (1861), 3d ed. (New York: Burt Franklin, 1969), pp. 19, 366–68. See also pp. 16, 40–41, 331, 369 –70.

21. Karl Marx, *The Poverty of Philosophy* (1847), in *Collected Works*, by Karl Marx and Frederick Engels (London: Lawrence and Wishart, 1977), pp. 207, 393–94. See Jon Elster, *Explaining Technical Change* (Cambridge: Cambridge University Press, 1983), pp. 163–71.

22. David Dickson, *The Politics of Alternative Technology* (New York: Universe, 1975), pp. 72–73, 181–82.

23. A. Bhaduri, "A Study of Agricultural Backwardness under Semi-Feudalism," *Economic Journal* 83 (1973): 120–37. See also S. Marglin, "What Do Bosses Do?" in Gorz, *The Division of Labor*.

24. Harry Braverman, *Labor and Monopoly Capital* (New York: Monthly Review Press, 1974), pp. 193–94, 199, 227–28. See also p. 188.

25. Ibid., pp. 196–206.

26. David F. Noble, *Forces of Production: A Social History of Industrial Automation* (New York: Alfred J. Knopf, 1984). Cf. James Fallows, "A Parable of Automation," *New York Review of Books* 31, no. 14 (27 September 1984): 11–17.

27. See Andrew Zimbalist, ed. *Case Studies on the Labour Process* (New York: Monthly Review Press, 1979); Dan Clawson, *Bureaucracy and the Labor Process:*

The Transformation of U.S. Industry, 1860–1920 (New York: Monthly Review Press, 1980). Also instructive in this regard is the earlier work of George E. Barnett, *Chapters on Machinery and Labor* (1926) (Carbondale and Edwardsville: Southern Illinois University Press; London and Amsterdam: Feffer and Simons, 1969).

28. Ibid., p. 169.

29. Ure, *Philosophy of Manufacturers*, p. 23.

30. See Edmund F. Byrne, "Robots and the Future of Work," in *The World of Work*, ed. Howard F. Didsbury, Jr., (Bethesda, Md.: World Future Society, 1983), pp. 30–38.

31. Dickson, *The Politics of Alternative Technology*, pp. 72–73, 181–82. Compare Ure's views on the need for machines to outdo foreign competition, in *Philosophy of Manufacturers*, pp. 31–32, 329.

32. Ure, *Philosophy of Manufacturers*, p. 20. See also pp. 1, 20–21, 23.

33. Lawrence B. Evans, "Industrial Uses of the Microprocessor," in *The Microelectronics Revolution*, ed. Tom Forester (Oxford: Basil Blackwell, 1980), p. 144.

34. E. Janicki, "Is There a Robot in Your Future?" *The Indianapolis Star Magazine*, 22 November 1981, p. 55.

35. See Arthur W. Burks and Alice R. Burks, "The ENIAC: First General-Purpose Electronic Computer," *Annals of the History of Computing* 3 (October 1981): 332–36, 386–88.

36. *The Impacts of Robotics on the Workforce and Workplace* (Pittsburgh: Carnegie-Mellon University, 14 June 1981), p. 52.

37. The following analysis is derived from Elster, *Explaining Technical Change*, pp. 96–111.

38. Robert Theobald, "Toward Full Unemployment," in *The World of Work*, p. 54. For Frithjof Bergmann's view, see, for example, "The Future of Work," *Praxis International* 3 (October 1983): 308–23.

39. Dickson, *The Politics of Alternative Technology*, pp. 55–56 and, in general, 41–62. See also Bernard Gendron, *Technology and the Human Condition* (New York: St. Martin's Press, 1977).

40. Braverman, *Labor and Monopoly Capital*, pp. 15–16, 22, 24.

41. Herbert Marcuse, *One-Dimensional Man*, (Boston: Beacon, 1966), pp. 35–37. Cf. pp. 44–45, 59, 231–32, 235. Marcuse bases his view on a passage from Marx's *Grundrisse der Kritik der politischen Oekonomie* in which Marx declares that labor time will eventually cease to be the measure of wealth.

42. Andre Gorz, *Paths to Paradise: On the Liberation from Work*, trans. M. Imrie (London: Pluto Press, 1985).

43. Adam Schaff explains Marx's early dreams of an "end of labor" as "youthful folly" categorically rejected in *Capital*. According to Schaff, "utopian prophecies" about what automation might accomplish "do not take us a single step further in the organization of our life today." Adam Schaff, *Marxism and the Human Individual*, trans. Olgierd Wojtasiewicz, ed. Robert S. Cohen (New York: McGraw-Hill, 1970), pp. 124–26, 134–35.

44. Sigmund Freud, *Civilization and Its Discontents* (1930), trans. James Strachey (New York: W. W. Norton, 1962), p. 2n. See Georges Friedmann, *The Anatomy of Work*, trans. Wyatt Rawson (New York: Free Press, 1964), p. 126; Philip Rieff, *Freud: The Mind of the Moralist*, 3d ed. (Chicago: University of Chicago, 1979), p. 245.

45. Eric Fromm, *The Sane Society* (New York: Rinehart, 1955), pp. 288–89, quoted by Friedmann, *The Anatomy of Work*, pp. 54–55.

46. Manuel and Manuel, *Utopian Though in the Western World*, pp. 745–47, 769. Note in particular authors' comments about the nineteenth-century debate regarding the value of work, p.745.

47. E. F. Schumacher, "Social and Economic Problems Calling for the Development of Intermediate Technology," mimeographed (undated), quoted by Dickson, *The Politics of Alternative Technology*, p. 153.

48. See David Macarov, *Work and Welfare: The Unholy Alliance* (Beverly Hills, Calif.: Sage, 1980); Frances Fox Piven and Richard A. Cloward, *Regulating the Poor: The Functions of Public Welfare* (New York: Vintage Books, 1972).

The Alienation of Common Praxis: Sartre's *Critique* and the Weberian Theory of Bureaucracy

FREDERIC L. BENDER

In this essay I shall refer to a considerable amount of empirical research on bureaucracy, perhaps more than most philosophers would find congenial, in the belief that one of the main tasks of an emancipatory social philosophy is the demystification of ideologies that justify and perpetuate social domination. As Mihailo Markovic has put it:

> When the philosopher asks those tough questions about the deeper meaning behind the successes of technology and improvements of material well-being, about the senseless waste of human and natural resources, about crippled existences, unnecessary suffering, ignorance and boredom, about still widespread material and spiritual misery, about the destruction of communal solidarity, about the possibilities and the ways of transcending the human condition—then he becomes dangerous, subversive. Surely he is dangerous for the existing power because he sees beyond it, demystifies it, undermines its most reliable weapon for manipulating people and ruling comfortably—its ideology.[1]

In the spirit of Markovic's comment, the structure of this essay is as follows. First, I discuss some of the main features of the Weberian analysis of bureaucratic authority. Second, I show how Weber's scientific account was transformed into an ideology legitimating a considerable amount of practice in bureaucratic capitalist societies. Unfortunately, space does not permit dealing with bureaucratic centralist ("actually existing socialist") societies here, so my analysis in this section is confined to detailing how the "scientific management" of Frederick Taylor and the "human relations psychology" of Elton

Mayo, and their respective followers, transformed the Weberian ideal typical model into an ideology. Third, I discuss briefly the contributions of contemporary critical theory to our understanding of bureaucracy and its ideological mystification. Fourth, I argue that critical theory, despite its contributions, is inadequate as a theoretical basis for revealing the reasons why bureaucracy is so fundamentally "against the grain," so basically dehumanizing, so widely hated despite its utility. It is chiefly in this last section that I shall draw upon existentialist philosophers, especially Sartre. Bureaucratic authority, especially in economic organizations, is an issue not altogether foreign to the concerns of philosophers. Mill argued that worker participation is an essential ingredient for the development of democratic values, and Marx asserted that industrial democracy would evolve only under socialism. From both these perspectives, the professional and technocratic justifications of managerial power can be seen to be primarily ideological rationalizations designed to advance and protect the interests of the administrative stratum.[2] And, as Alisdair MacIntyre recently has argued, the very methodology of organization theory has become the ideology of bureaucratic authority—little more than an "ideological expression of that same organizational life which the[se] theorists are attempting to describe."[3]

THE WEBERIAN MODEL

Weber's main objective was to describe and legitimate the replacement of "old-world" forms of state authority, based on traditional and charismatic leadership, with a "rational-legal" mode of authority that facilitated industrial development and was adequate to the rigors of capitalist competition. Traditional and charismatic forms of authority stood in the way of profit-maximizing and were too slow in responding to rapidly changing events and modern communications. The charismatic mode of leadership was relatively inefficient because of its reliance on emotion and mystique instead of fixed rules and routines. In rational-legal authority structures, the obedience and compliance of subordinates would be based on adherence to rules rather than on relations to particular persons. Even superiors would have to comply with regulations, thus anchoring authority to the rules governing the rational pursuit of goals.[4] For Weber, the character of bureaucracy is the same whether it is private or public. His idealized model of bureaucracy as the quintessential form of legal-rationalistic authority emerged as the basic conceptual foundation of organizational analysis. Subsequent theory, however, has underplayed Weber's emphasis on rational-legal authority as a form of *power*, stressing instead its *technical-efficient* aspects.

Weber argued that the decisive reason for the advance of bureaucratic organization is its purely technical superiority over any other form of organization. Work organized by collegiate bodies causes friction and delay, requiring compromises between conflicting interests and personalities. Administration therefore runs less precisely and is more independent of rules, hence is less unified and slower.

As Weber accurately foresaw, today it is primarily the capitalist market economy that demands that the administration be discharged precisely, unambiguously, continuously, and with as much speed as possible. The extraordinary increase in the speed in which public announcements, as well as economic and political facts, are transmitted exerts a steady and sharp pressure in the direction of speeding up administrative reaction. Weber believed that the optimum reaction time is normally attained only by a strictly bureaucratic organization.

Bureaucratization offers above all the optimum possibility for carrying out the specialization of administrative functions according to purely objective considerations. The "objective" discharge of business primarily means a discharge of business according to calculable rules and "without regard for persons." Persons, whether within the organization (at any rank) or among its clients, must adjust to the organization, not the other way about.

Weber emphasized that modern culture, specifically its technical and economic basis, demands calculability of results. Such calculability, which is welcomed by capitalism, develops the more perfectly the more the bureaucracy is depersonalized, the more completely it succeeds in eliminating from official business love, hatred, and other personal, irrational, and emotional elements that otherwise might escape calculation. Also, the more complicated and specialized modern culture becomes, the more its apparatus demands the personally detached and strictly "objective" expert.[5] Bureaucratic structure goes hand in hand with the concentration of the material means of management. According to Weber, once it is fully established, bureaucracy is among those social structures that are the hardest to destroy. Under otherwise equal conditions, a societal action that is methodically ordered and led is superior to every resistance of individual or common action. Where the bureaucratization of administration has been completely carried through, a form of power is established that is practically unshatterable.

Weber also pointed out that the ruled increasingly cannot dispense with or replace bureaucratic authority once it exists, for bureaucracy rests upon expert training, a functional specialization of work, and an attitude set for habitual and virtuoso-like mastery of single, yet methodically integrated, functions. Increasingly, the material fate of the masses depends upon the steady and correct functioning of the bureaucratic organizations of private capitalism and the state, and the idea of eliminating these organizations becomes more and more utopian. Likewise, in Weber's view it is naive to believe that bureaucratic organization can be destroyed, for such a notion overlooks modern man's desire to keep to habitual rules and regulations, now that such compliance has been conditioned into the governed as well as into the officials.[6]

Yet we must be on guard against the tendency to reify bureaucracy and see it as a socially and historically independent variable. Rather, as Weber himself pointed out, although the modern state is undergoing bureaucratization everywhere, it remains an open question whether the bureaucracy's power within the polity is universal, for bureaucracy's ever-increasing "indispensability" is no more decisive for this question than the economic indispensability of slave labor was in slave societies. Whether the power of bureaucracy increases *as such* cannot be decided a priori. However, under normal modern conditions, a fully developed bureaucracy always proves overwhelming. The "political master" finds himself in the position of the dilettante who stands opposite the "expert," regardless whether the master is an aristocratic or collegiate body, legally or actually based on self-recruitment, or a popularly elected president or hereditary, absolute, or constitutional monarch.

THE IDEOLOGICAL TRANSFORMATION OF THE WEBERIAN MODEL IN BUREAUCRATIC CAPITALISM

In this section I want to distinguish two successive transformations of the Weberian model: the "scientific management" school (Taylorism) and the "human relations" movement (especially the work of Elton Mayo). As is well known, Weber's conception of bureaucracy as a model of efficiency converges with the contemporaneously emerging "value-neutral" science of management, which seeks to uncover the maximally efficient rules governing the rational pursuit of organizational goals. Shunning Weber's broad political perspective on entrepreneurial legitimation, his concept of authority was restricted largely to an emphasis on its functional contribution to efficient organizational practices, such as the use of hierarchy, technical competence, and leadership,[7] making Weber's highly political concepts appear to be value neutral and purely technical, in effect reflecting the sort of authority he had described.

Now, whereas Weber had emphasized the problem of authority, Frederick Winslow Taylor focused on the advantages of the bureaucratic reorganization of the production process itself, recognizing that capitalists could never win their struggle with labor under the divided authority that characterized the nineteenth-century workplace. As long as capitalists continued to take for granted that workers' craft organizations would retain control of the details of the labor process, they were forced to depend upon their voluntary cooperation and initiative.

The functions of Taylor's scientific management experts were twofold. On the one hand, they were to enter the workplace to learn, through time and motion studies, what the workers already knew: how to plan and carry out the details of the work process. On the other hand, through managerial analysis and planning, Taylorites were to employ this knowledge to redesign "efficiently" the production process under management's control. Taylorism's real contribution was less the development of scientific techniques for measuring work processes[8] than the construction of a new mode of organizational control, wresting workplace authority from the craft organizations and placing it in the hands of the newly emerging professional managers in the great labor struggles just prior to and during World War I.

By the mid-twenties, however, it had become clear that, from management's viewpoint, although the Taylorites had properly identified the issue of workplace authority, they had not found quite the right mechanism for sophisticated workplace control.[9] In this light, the success of the human relations movement can be understood as the culmination of a series of attempts to find that "right" mechanism of control, in response to the informal work groups and upsurge of organized labor stimulated in considerable part by hostility toward Taylorism itself.[10] Mayo and his followers accepted scientific management's bureaucratic model of the workplace but developed their "human relations" school of organization theory in response to the limiations imposed by Taylor's overly physical orientation to work, in the process introducing behavioral science techniques and the idea of the "informal social group." In practice this led to the development of manipulative techniques to be employed by personnel managers.

Mayo's operating assumption was that modern organizations

could be made reasonably, if not ideally, rational in Weber's sense, by replacing their "irrationality" with formally rational actions carried out under management control.[11] As unabashed students of industrial efficiency and stable work relations, human relations' first theorists laid the groundwork for a discipline designed to supplement and support the bureaucratic mode of authority and control. The outcome was an organization theory that has "by its assumptions and procedures systematically narrowed its field of inquiry and, consequently, presented an incomplete and distorted picture of organizational reality,"[12] thereby functioning as an ideology in both the descriptive and pejorative meanings of the term. In the pejorative sense, an ideology systematically deludes some or all of the agents in a society about themselves, their social position, or their interests. It is by virtue of the fact that it supports or justifies relations of repressive authority, exploitation or domination, that a form of consciousness is an ideology. Organizational ideology can be understood as an attempt to justify the leaders' privilege of voluntary action and association while imposing upon their subordinates the duties of obedience and service to the best of their ability.[13] Thus, the human relations movement, cloaked in the garb of scientific rationality, has translated images of repressive authority into widely accepted forms of workplace consciousness.[14]

Mayo's most important finding was his recognition of the importance of the supervisory climate. His researchers sought to reeducate shop-floor supervisors by teaching them to become well liked by their subordinates, and to prove the importance of personnel counselors, who were to try to take the place of the informal groups, although, when dealing with workers' attitudes toward their problems, they were not to deal with the problems themselves. They were to listen to any problem of any employee, but neither to give advice nor argue, nor to make promises about remedial action. They were simply to watch for signs of unrest and try to assuage the tension of the worker verbally before the unrest could spread and become active. Counselors were to try to dilute or redirect dissatisfaction by helping the employees to think along "constructive" lines, in effect adjusting people to situations, rather than situations to people. Mayo's work opened up new vistas for supervision, once the disruptive potential of the informal group could be neutralized. Management could train supervisors to establish a harmonious work climate, free of idiosyncratic personal authority. This link between supervision, morale, and productivity became the foundation of academic human relations theory.[15]

Mayo also maintained that the cooperation of individuals and groups is the supreme moral principle. Where there are different group interests, even certain inevitable conflicts, their elimination is merely a matter of "intelligent organization that takes careful account of all the group interests involved."[16] In his view, industrial society, as presently organized, leads to the social maladjustment of workers and eventually to obsessively irrational behavior, including the formation of adversarial unions. Social class conflict is thus a deviation from normal human actions and attitudes, little more than a primitive expression of human imperfections, the remedy for which is the proper application of psychological techniques and the development of "social skills" education, which would eliminate the need for union representation altogether.[17]

Management, in Mayo's view, is the embodiment of logic and rationality, and hence always knows best. Its interests are synonymous with the interests of the organization and of society as a whole. Thus, not only does human relations theory justify the dominance of management over labor, it also provides psychological techniques for blurring workers' consciousness of the realities of workplace control. It posits the logical, rational authority of management over the irrational, psychologically immature behavior of workers and their unions. Management is the agent of cooperation, whereas unions embody economic and social conflict. In psychological terms, management exemplifies the all-knowing, benevolent father who offers guidance and protection to his children, the worker's. Although this is never stated formally the result is to blur the worker's consciousness of the general issues of power, authority and class, particular unjust practices, and repression.[18]

THE CRITICAL THEORY OF BUREAUCRACY

Ignoring the classic struggles between labor and management that gave shape to modern bureaucratic structures, mainstream writers have emphasized the technical imperatives of industrialization in explaining the development of bureaucratic forms of organization. Managers, in view of their technical and administrative training, emerge from such an analysis as the most valuable members of the organizational system; for only managers, not workers, can plan and oversee the technical complexity of modern organizational processes.

In recent years, as the ideological effects of scientific management and human relations theory have become increasingly clear, a critical theory of bureaucracy has arisen. For the critical theorist, the failure of conventional theory is its unwillingness to look beyond the managerial point of view, no doubt facilitated by management's ability to make substantial contributions to organizational research and by the political and cultural biases favoring management in capitalist society. Failing to see more than one logic of rationality in the workplace, mainstream theory confuses management's interests and motives with those of the organization itself, its members, its clients, and society in general. The attempt to build a critical organization theory can be understood as an attempt to offset this narrow efficiency-oriented conception of organizational rationality, and as a search for a broader conception. Unlike mainstream theory, critical theory emphasizes the elitist character of modern organizations, draws attention to issues of social class conflict, focuses on power as the primary currency of organization dynamics, stresses the social and historical context of the development of organizations, and favors nonpositivistic methodologies. Antecedent to the problem of efficient work, organizations must be conceptualized as tools for the pursuit of personal, group, or class interests within the large framework of sociopolitical interest.[19]

Organizational work turns out to be actually two distinct types of labor performed simultaneously: political and productive labor.[20] Most mainstream organizational research has emphasized only the analysis of the production-related dimensions of work. Political labor, on the other hand, is the primary focus of critical organizational theorists. Directed by a managerial elite, the administrative structure is an internal political hierarchy. The function of the administrative system is to set the conditions under which production is appropriated, controlled, and distributed. As Braverman has argued, production in capitalist organizations is never simply organized to conform to the logic of efficiency and technology; organization follows whatever path best allows management to prevent workers from controlling their own work, and/or the distribution of production.[21]

Conventional theorists, of course, recognize the existence of social class differences, but they do not confront these on their own terms. Goldman states, "underneath this recognition of organizational conflicts, most theorists assume that the fundamental interests of workers and managers are neither contradictory nor antagonistic."[22] Without a critical class analysis, conflicts between owners/managers and workers/unions are generally misinterpreted as dysfunctional. In this light, it is difficult to see that the resolution of many of these class conflicts may lie in the transformation of the social relations of work rather than simply by minor bureaucratic modifications.[23]

For mainstream theory, management's function is primarily to monitor and facilitate the sociotechnical adjustments dictated by technological change. Organizational problems are conceptualized as atemporal and unchanging technical-administrative matters. The lack of historical perspective freezes analysis into the present and reifies the goals of the *status quo*. In this regard, organizational goals often appear to emerge from the organizations themselves rather than from the critical decisions of the dominant coalition or managments that structure and control the organization.[24] A proper respect for context requires that we reject the traditional mainstream emphasis on finding alleged value-neutral, universal laws of organizational structure. The notion of such universal generalizations assumed in the mainstream (especially systems theorists') search for a general theory of organizations is borrowed from an outdated conception of the natural sciences. At the basis of these assumed causal relationships is the presupposition that there is one best structure for all organizations and one best way for them to function. Not only is such a mechanistic perspective normatively problematic, its statistical correlations tell us almost nothing about why organizational phenomena occur as they do.

EXISTENTIALISM: BEYOND THE CRITICAL THEORY OF BUREAUCRACY

To the extent that bureaucratic domination or manipulation are indeed unjustified, critical theory certainly has an important contribution to make. We are still left, however, with two unanswered questions: Why, as Weber had already shown, do so many people within bureaucracy identify their interests with it? Why, as the clients or customers of bureaucracies, do they remain indifferent to its dehumanizing effects? It is the merit of the existentialist approach to the human condition to reveal the central importance of the sense of selfhood and to suggest that, to the extent that bureaucracy cuts deeply into the sense of selfhood, the critical theorists' consideraions of empowerment or its lack, for all their insight, conceal the most basic reason for bureaucracy's power. Contra critical theory, it is not the case that a person is nothing but bureaucracy's powerless victim in his or her role as worker or consumer. On the contrary, a person is before everything else a subject of experience and action, a "project," or a "will to meaning." It is the merit of the existentialist approach to make clear that it is precisely this unique, individual subjectivity which bureaucracy, its ideologization, and the critical theory thereof, all alienate from our understanding, each reducing the concrete individual to a "case," a "number," or a member of a certain class, respectively, without significant remainder.

Again, I certainly have no quarrel with critical theory's assumption that people need to control the institutions (whether economic, political or otherwise) that effect their daily lives. And surely bureaucracy is an ever-present obstacle to the satisfaction of this need. But if by some miracle bureaucratic organizations were replaced by collegial or cooperative ones, the existential basis of sustaining common praxis in a nonbureaucratic manner—and maintaining this autonomy over time—would still be lacking, for the "bureaucratization of the world"[25] means, among other things, the destruction of the personal expectation of, and capacity for, nonbureaucratic action and organization. This is why empowerment (the elimination of bureaucratic structures) can be only a necessary, not a sufficient, condition of sustained resistance to bureaucratization and of the creation of debureaucratized society. In this section I shall analyze how bureaucracy alienates us from our sense of self, an alienation that is typically overlooked by critical theorists because of their exclusively sociologistic or economistic perspective. Despite the many well-known differences among its leading representatives, existentialism, if it is anything, is a particularly determined attempt to speak out against historical tendencies and social institutions that destroy the integrity of the individual as a subject. Note that, with few exceptions, the existentialists' understanding of the individual differs sharply from the acquisitive, atomistic individualism of classical liberal theory and its successors. There is no significant twentieth-century existential philosopher who regards acquisitiveness or autarchy as representative of an "authentic" freedom or the fulfillment of the human *telos* as they understand it. To the extent that some existential philosophers have chosen to emphasize individuality more than sociality (*Mitsein*), or whose work lends itself to such an interpretation, they too have contributed to a distorted picture of the human condition. The later Sartre, as we shall see, goes quite far toward explicating existentialism as a *social* philosophy relevant, among other things, to the analysis of bureaucratization and the struggles against it.

The classic existentialist distinction between human existence and that of other entities is that, unlike the latter, which simply have their existence as what they are, the human individual must *make* his or her own existence at every moment: a person's extranatural aspect is not there from the outset but is an aspiration or a project. This project is what we feel to be our true being; it is what we call our personality or self. A person is fundamentally not what he or she has been or is already, but what he or she is not yet; for at every moment I face diverse possibilities, among which I must choose. In Sartre's words, "man is nothing else than that which he makes of himself."[26] Of course, I invent my projects in the light of my circumstances or my "facticity," including my embodiment, my past, my occupation, my inevitable but indeterminate death, the underlying social fact of scarcity, and so on. However, within this facticity, I am to some extent free to define what in my circumstances I will mold into *my* "situation." In short, an individual is finitely self-creative: one is what one has done with what has happened to oneself.

An individual is thus defined in terms of his or her passion and its mastery, independence of convention, and that creative freedom that finds ultimate expression in being a law unto onself. On Sartre's analysis, self-deception, or "bad faith," consists in the delusion that one has become some sort of a *thing*, has fixed oneself once and for all, no longer needs to make oneself, and no longer is responsible for what one has become. On the contrary, Sartre avers, our "condemnation" to freedom can be illustrated by the fact that a person cannot simply be a homosexual or a waiter, for example, in the same way in which he happens to be six feet tall or blond, because he cannot *not* choose continuously among his outstanding possibilities, including whether to continue to be a homosexual or a waiter.[27]

But an individual's situation is always to be among others in various types of relationship, of which bureaucracy is one. Bureaucracy views its functionaries and clients alike as things, as what we would be if we were to be permanently in self-deception as Sartre conceives it. Rational-legal authority assumes individuals are the polar opposite of self-creative subjects: a worker's or client's project is totally irrelevant to bureaucracy's formal, technical, and supposedly value-neutral, operation. In the decision process within a bureaucratic organization, rules and precision take precedence over emotion; calculability of results takes precedence over regard for persons. Indeed, specialization according to functional efficiency reinforces the alienation of the bureaucrat himself, because his fragmented job can have little relation to his fundamental project, unless, of course, he has indentified the two, in which case his project becomes simply that of moving up the hierarchy.[28]

The increasing prevalence of bureaucratic institutions in contemporary society forces us to put on a "face" that will enable us to succeed in encounters with such institutions: we are required, in effect, to lie-for-others.[29] Within bureaucratic institutions, however, functionaries find themselves in constant danger of coming to believe their own lies and losing their sense of self entirely. Once I have absorbed the external standard imposed by the bureaucratic organization within which I work, I cease to have any unique being-for-myself. Presumably, this is why bureaucrats, although they know they suffer, feel forced to return to the institution again and again, for, outside their offices they fear that they have no identity at all.[30]

From the perspective of bureaucracy's clients, the functionary acts according to mysterious norms, deprived of the time and even the right to understand the client's individual needs. Communication takes place in a one-sided manner, with the bureaucrat having the advantage of availing himself of a jargon he knows is unfamiliar to the client. Indeed, the client, who uses merely ordinary language, can never be sure that the bureaucrat understands the context and meaning of his or her assertions. In short, bureaucracy transforms subjects (individuals as self-creative projects) into objects. As a result, bureaucracy is incompatible with the dignity of man and with the freedom of man as a social *and* individual being, as we shall see now in the later Sartre's existential Marxist theory of bureaucracy as the alienation of common, or social, praxis.

THE ALIENATION OF COMMON PRAXIS (SARTRE'S THEORY OF BUREAUCRACY)

As noted previously, our individuality or sense of self as project is but one aspect of selfhood. The second is that of our *inter*subjectivity or "being-with-others." In the words of the early Sartre:

> The other is indispensable to my existence, and equally so to any knowledge I can have of myself. The intimate discovery of myself is at the same time the revelation of the other as a freedom which confronts mine, and which cannot think or will without doing so either for or against me. Thus, at once, we find ourselves in a world which is that of "inter-subjectivity." It is in this world that man has to decide what he is and what others are.[31]

Because man is a project that possesses a subjective life, man is responsible for what he is. Thus Sartre places the entire responsibility for his existence squarely on his own shoulders. But Sartrean individuality is not individualistic:

We do not mean that he is responsible for only his own individuality, but that he is responsible for all men. In choosing for himself he chooses for all men. There is no action that is not at the same time creative of an image of man such as he believes he ought to be. Our responsibility is thus much greater than we had supposed, for it concerns mankind as a whole. My action is, in consequence, a commitment on behalf of mankind. I am thus responsible for myself and for all men, and I am creating a certain image of man as I would have him to be. In fashioning myself I fashion man.[32]

This has as a consequence an ethic of freedom incompatible with any radical individualism:

Freedom, in respect of concrete circumstances, can have no other end and aim but itself; and when once a man has seen that values depend upon himself, . . . he can will only one thing, and that is freedom as the foundation of all values. . . . And in thus willing freedom, we discover that it depends entirely upon the freedom of others and that the freedom of others depends upon our own. I cannot make liberty my aim unless I make that of others equally my aim.[33]

Nonetheless, Sartre came to realize that the Husserlian framework of *Being and Nothingness* was inherently oriented toward emphasizing the perspective of the self in isolation from the other. Despite Sartre's own efforts, subjectivity on this view must remain private, for the other is always given primarily in perception, as a body, as an object.[34] Although it is possible to describe *inter*-subjectivity, or self-other relations, in this framework, it is impossible to describe *co*subjectivity, where a common or group project constitutes the subjectivity of its participants. Sartre's subsequent realization of the inadequacy of his earlier ontological framework led eventually to the existential Marxism of his *Critique of Dialectical Reason*, which he announced as follows:

I consider Marxism the one philosophy of our time which we cannot go beyond and . . . I hold the ideology of existence . . . to be an enclave inside Marxism, which simultaneously engenders it and rejects it. . . .

We are convinced at one and the same time that historical materialism furnished the only valid interpretation of history and that existentialism remained the only concrete approach to reality. . . . Existentialism has been able to return and to maintain itself because it reaffirmed the reality of men as Kierkegaard asserted his own reality against Hegel. . . .

As soon as there will exist for everyone a margin of real freedom beyond the production of life, Marxism will have lived out its span; a philosophy of freedom will take its place.[35]

For Sartre, the existential "enclave" within Marxism is necessary because orthodox Marxists had suppressed the moment of individual praxis in favor of the quasideterminism of the alleged laws of history and its alleged economic "base."[36] Existential Marxism is to correct this bias by capturing the individual in the act of living his alienation.[37] Within the framework of the history in which we are all actors, a history dominated by the fact of scarcity, being-for-itself is made concrete by being replaced by praxis and being-in-itself by inert matter. Sartre now recognizes that an atmosphere of violence, of class conflict—whether overt or suppressed—must remain a basic fact of existence as long as scarcity continues to prevail. This decisively alters his concept of the individual, his facticity and his project.

In the *Critique* man is considered first as abstract, individual praxis, "going beyond material conditions towards an objective, and imprinting itself through labor in inorganic matter as a reworking of the practical field."[38] Praxis originates in need, thereby impelling the reworking of matter, in the process of which man establishes historically specific relationships of reciprocity. But in a social world defined by scarcity, such relationships inevitably entail domination and subordination. Originally free activity (in the ontological, not temporal, sense) becomes passive and inert, like matter—a condition Sartre calls "seriality"—the form of domination of which modern bureaucracy is a prime example. In serial relations, men regard structures as given, externally imposed, and make no effort to alter them. In contrast to the series, however, there is the possibility of a type of social organization Sartre calls the "group," which can come into being through that common praxis by which (social) man "totalizes," that is, acts to achieve a certain common objective, thereby transcending serial relations by internalizing the common goal or purpose unanimously accepted by all concerned. In such a "group-in-fusion," no hierarchy or leader is required. Unfortunately, according to Sartre, the dialectical movement from series to group can be reversed in various ways, the most important of which involves the group's bureaucratization and its transformation into an "institution."[39]

If relations in the series resemble the inertness of matter, the group-in-fusion on the contrary possesses the vitality of a free (common) project. Groups are formed in reaction to the oppressive inhumanity of the series and the impossibility of free beings living in such inert structures, in the process replacing mechanical with organic solidarity. The group is formed through a reciprocity of praxis where each sees in the other the same project as his own: each individual reacts "in a new way: not as an individual, nor as an Other, but as an individual incarnation of the common person."[40] Such traits of the series as impersonality, isolation, and atomization disappear as the intensely personal relations of the "group-in-fusion" are constituted. For Sartre, the group is "the beginning of humanity," where men have "recuperated their lost being," their suppressed freedom.[41]

Poster has pointed out that, in arguing that the recovery of freedom is possible only in the group, Sartre breaks decisively from the liberal conception of individual freedom. Whereas the individualist tradition found autonomy only in the atomized individual, the *Critique* argues that freedom is not centered on the individual self, but requires a thoroughly democratic (or anarchist) group structure, in which each person is recognized as a free being sharing in the common action directed against the serial structures of domination that atomize and alienate him or her. Unlike liberalism's ideal discussion group in which agreement is sought through open debate with complete toleration for verbal self-expression, experience in the group-in-fusion is total, engaging the full, concrete existence of the individual, not just his or her rationality.[42]

Within serial relations such as bureaucracy, freedom is sharply limited, for in such a condition people cannot recognize and confirm the freedom of others due to their mutual indifference or antagonism. As a result, free action within seriality is not directed against alienation and its causes, but toward "success" within the serialized institution. In the group-in-fusion, on the other hand, freedom becomes both end and means, constituting the social milieu as *human*. Sartre recognized that, without a proper material base, the group-in-fusion must remain a fleeting and unstable structure, although, in his polemics against ossified, orthodox Marxism, he argued that any movement toward a free society requires the mediation of the democratic group-in-fusion, not the elite Leninist party, as its proper revolutionary organization.[43]

However, Sartre does not romanticize the group as if it were easy to realize and maintain. He introduces a somber note in arguing that the group-in-fusion faces danger on two sides, however, launching it onto the dialectical path toward its own undoing. On the one hand, it constantly risks petrification into seriality. On the other, it is threatened by the very freedom of its members from which it was constituted, for they can always withdraw from it. The same free praxis by which the group was formed can become its mortal enemy. The group must thus inaugurate an internal "reign of terror" in order to suppress its possible destruction from within. As a result, Sartre argues that the group's destruction is inexorable: because it requires a certain degree of job specialization, it eventually becomes an "organization," with a distribution of well-defined functions. Although some spontaneity of action is lost at this stage, the separation of tasks as such does not necessarily destroy the group's unity. Yet, through specialization, the group increasingly slides into the very condition that it was created to negate. When it finally becomes an institution it is in essence indistinguishable from the series it originally was formed in protest against.[44]

As the group-in-fusion is transformed into an organization and thence into an institution, from the individual member's perspective, the common purpose becomes increasingly subordinated to his narrow obligation to perform his tasks, thereby "exteriorizing" his praxis. Each member's originally freely chosen identification with the group becomes more and more like an alien obligation. As people identify primarily with their own roles, their interactions become less self-directed and more based on external reciprocity. Authority must develop to insure that each member performs his task appropriately. This growing serial quality of the institution corresponds to the individual's increased loss of power; the individual can no longer recognize his free project in the now organized division of labor. The legitimate and necessary functions performed by leaders become sources of alienation for the followers, representing to them a profound degradation of the group as common praxis, preserving the group's structural bond only under a stupefying form.[45] Because loosely defined groups cannot survive in a practical field pervaded by the series, all sense of intragroup spontaneity is lost, hence authority is needed to assure the group's unity of purpose, reconstituting the external reciprocity of the series.[46]

CONCLUSION: THE WEBERIAN AND SARTREAN MODELS OF BUREAUCRATIC CAPITALISM

If it can be taken as established that the Weberian model is incipiently an ideological defense of class domination (becoming explicitly so in scientific management, human relations theory, and the work of mainstream organization theorists—not to mention its appropriation in the Leninist/Stalinist context), the question arises whether Sartre's analysis is also ideological, also a form of *false* consciousness, systematically deluding the victims of bureaucracy while serving to legitimate structures of domination.

This dialectic of series and group suggests that in Sartre's view the *tendency* towardbureaucracy is perpetually present, although groups become bureaucratized at varying rates and new groups are potentially always capable of being fused in opposition to established institutions. Thus we search the *Critique* in vain for an affirmation that bureaucracy, or the danger of bureaucratization of even the most freely constituted group, is likely to disappear. As in his earlier works, with their more individually oriented analyses, Sartre's view would seem to be that it is only in this constant possibility of struggle, this possible totalizing praxis of individuals fusing themselves into groups expressive of their aspirations for freedom, that our freedom—and hence our responsibility, as beings of praxis—is to be located.

In the *Critique* Sartre characterizes bureaucracy as "a total suppression of the human," except at a minute point at the top of the hierarchy, which is supported by the impotence of the masses at the bottom. Sartre thereby implicates the masses in their domination, manipulation, and "mineralization by bureaucracy,"[47] for to some extent they have abrogated their own freedom.

As we have seen, because bureaucracy's defining characteristics are mistrust, inertia, the suppression of initiative, and the dissolution of identity and its reconstitution only through relation to a higher authority,[48] Sartre opposes the "scintillating" life of the fused group, which is the highest fulfillment of our sociality, to the "old worm-eaten sovereignty of the bureaucracy."[49] But he reaffirms, in the midst of all this, the ever-present possibility of group differentiation and regroupment from below—the possibility of antibureaucratic struggle emerging spontaneously and likewise the possibility of continued struggle to maintain free praxis within groups. Seen in this light, Sartre has moved the question of bureaucracy's (and our) future from the naive issue of whether bureaucracy can somehow be eliminated, to the more realistic, albeit less romantic one of raising anew the issue of the responsiblility of social individuals for their own fate—and for the fate of societies, institutions, organizations, and groups to which they belong. If, on Sartre's analysis, bureaucracy cannot be miraculously banished, if Sartre offers us no model of nonbureaucratized institutions (which for him would be a contradiction in terms), he does offer us support for a sober understanding of our (that is, bureaucracy's victims') own responsibility not only for the existing state of affairs but also for that praxis necessary for forming the authentically, freely fused groups from within which a struggle, without illusions, against the bureaucratization of everyday life might take place. Clearly, the events of 1968 reflected aspirations similar to those Sartre had articulated in the *Critique*, as bureaucratization touched off massive student resistance througout the advanced capitalist world (and elsewhere, including many bureaucratic centralist societies)—a revolt among those who were about to become bureaucracy's favored recruits (many of whom subsequently became such).

I would suggest that the merits of Sartre's analysis are due primarily to his ability to fuse the critical theory of bureaucratic domination with the existential-Marxist dynamic of alienation. In conclusion, I would like to suggest that Sartre *does* offer an answer to the most fundamental question raised by bureaucracy: Why do people (functionaries as well as their clients) accept bureaucracy even though they hate it? Bureaucratic capitalism has prepared and conditioned us to accept seriality, not only by its promulgation as an ideology, and not only by bribing us with the material benefits— many of which are illusory—of its much-vaunted efficiency (mass consumption of goods and the benumbing spectacles of mass culture), but by creeping into our souls and creating the illusion of the normalcy of *being-no-one* outside of one's relation to the institution. What is possibly ideological in Sartre's account is his argument that bureaucracy arises inevitably from revolt against its own excesses, as the dialectic of freedom and authority is played out within groups (and whole societies). But this circularity need not be an inevitable "useless passion." Sartre's analysis of this dialectic holds out for us the possibility that the struggle for emancipation from bureaucracy can, indeed, spiral toward the outer limits of freedom possible in a world of scarcity. It is the merit of Sartre's existential-Marxist critique of bureaucracy to have shown us that, to the extent that we no longer know what we *want*, but only what we are *against*, there can never be a permanently successful revolution against bureaucratic domination. This analysis does *not* preclude an eventual overcoming of bureaucracy, even if it does argue against the possibility of a permanent obliteration of its possibility. For this reason, Sartre's analysis, unlike Weber's, is an important contribution to mankind's future emancipatory struggles, not an ideology.

NOTES

1. Mihailo Markovic, "Reason and Historical Praxis," in *Marxist Humanism and Praxis*, ed. Gerson S. Sher (Buffalo: Prometheus, 1978), p. 21.

2. Frank Fischer and Carmen Sirianni, ed., "Introduction," in *Critical Studies in Organization Theory and Bureaucracy* (Philadelphia: Temple University Press, 1984), pp. 13–14.

3. Alasdair MacIntyre, "Social Science Methodology as the Ideology of Bureaucratic Authority," in *Through the Looking Glass*, ed. Maria J. Falco (Washington: University Press of America, 1979), p. 42. Cited in Fischer and Sirianni, *Critical Studies in Organization Theory and Bureaucracy*, p. 173.

4. Fischer and Sirianni, "Introduction," in *Critical Studies in Organization Theory and Bureaucracy*, p. 7.

5. Max Weber, "Bureaucracy," in ibid., pp. 31–33.

6. Ibid., pp. 35–37.

7. Frank Fischer, "Ideology and Organization Theory," in ibid., p. 174.

8. Ibid., p. 177. Fischer points out that, while Taylorism was able to show productivity increases, there is no basis to determine whether these resulted from improved work procedures or merely speeding up existing ones.

9. Richard Edwards, *Contested Terrain* (New York: Basic Books, 1979), p. 104. Cited in *Critical Studies in Organization Theory and Bureaucracy*.

10. Fischer, "Ideology and Organization Theory," pp. 177–78.

11. Fischer and Sirianni, "Introduction," pp. 8–9.

12. Paul Goldman, "Sociologists and the Study of Bureaucracy: A Critique of Ideology and Practice," *The Insurgent Sociologist* 3 (Winter 1978): 21. Cited in Fischer, "Ideology and Organization Theory," p. 172.

13. Reinhard Bendix, cited in Fischer, "Ideology and Organization Theory," p. 173.

14. Fischer, "Ideology and Organization Theory," pp. 172–73.

15. Ibid., pp. 182–83.

16. Elton Mayo, *The Social Problems of an Industrial Civilization* (London: Routledge, 1949), p. 128. Cf. Fischer, "Ideology and Organization Theory," p. 183.

17. Cf. Fischer, "Ideology and Organization Theory," pp. 183–84.

18. Ibid., p. 184.

19. Fischer and Sirianni, "Introduction," pp. 10–11.

20. Cf. Randall Collins, *The Credential Society* (New York: Academic Press, 1979).

21. Fischer and Sirianni, "Introduction," p. 12.

22. Ibid., p. 15; Cf. Goldman, "Sociologists and the Study of Bureaucracy," p. 26.

23. Fischer and Sirianni, "Introduction," pp. 14–15.

24. Ibid., pp. 15–16.

25. Henry Jacoby, *The Bureaucratization of the World* (Berkeley: University of California Press, 1973).

26. Jean-Paul Sartre, "Existentialism is a Humanism," in *Existentialism from Dostoyevsky to Sartre*, ed. Walter Kaufmann (New York: New American Library, 1975), p. 349.

27. Jean-Paul Sartre, *Being and Nothingness*, trans. Hazel Barnes (New York: Washington Square Press, 1968), p. 567.

28. As was observed by the young Marx in 1843. *Critique of Hegel's "Philosophy of Right,"* ed. Joseph O'Malley (Cambridge: Cambridge University Press, 1970), pp. 46–48.

29. Ralph P. Hummel, *The Bureaucratic Experience* (New York: St. Martin's Press, 1982), p. 116.

30. Ibid., pp. 116–17.

31. Sartre, "Existentialism," pp. 360–62.

32. Ibid., pp. 349–50.

33. Ibid., pp. 365–67.

34. For an illustration of the difficulties of constituting intersubjectivity from an egological perspective, see Edmund Husserl, *Cartesian Meditations: An Introduction to Phenomenology*, trans. Dorian Cairns (The Hague: Martinus Nijhoff, 1960).

35. Sartre, "Search for a Method," in Kaufmann, *Existentialism from Dostoyevsky to Sartre*, pp. 369–74.

36. Cf. Jean-Paul Sartre, "Science and Dialectic," *Man and World* 9, no. 1 (February 1979): 60–74.

37. Mark Poster, *Existential Marxism in Postwar France* (Princeton: Princeton University Press, 1975), p. 285.

38. Cited by William Leon McBride, "Jean-Paul Sartre: Man, Freedom, and Praxis," in *Existential Philosophers: Kierkegaard to Merleau-Ponty*, ed. George Alfred Schrader (New York: McGraw-Hill, 1967), p. 277.

39. Ibid., pp. 299–300.

40. Jean-Paul Sartre, *Critique of Dialectical Reason*, ed. Jonathan Reé and trans. Alan Sheridan-Smith (London: New Left Books, 1976), p. 357.

41. Poster, *Existential Marxism in Postwar France*, pp. 288–89.

42. Ibid., pp. 289–91.

43. Sartre, *Critique of Dialectical Reason*, pp. 658–63.

44. Poster, *Existential Marxism in Postwar France*, pp. 291–93.

45. On guard against the tendency to multiply the functions of revolutionary government, Marx distinguished between those that are necessary and those that are superfluous in *The Civil War in France*. For a discussion of the significance of this distinction, cf. F. L. Bender, "The Ambiguities of Marx's Concepts of 'Proletarian Dictatorship' and 'Transition to Communism,'" *History of Political Thought* 2, no. 3 (November 1981): 544.

46. Poster, *Existential Marxism in Postwar France*, pp. 293–94.

47. Sartre, *Critique of Dialectical Reason*, p. 658.

48. Gila J. Hayim, *The Existential Sociology of Jean-Paul Sartre* (Amherst: University of Massachusetts Press, 1980), p. 134.

49. Sartre, *Critique of Dialectical Reason*, p. 659.

Artificial Representationalism: Necessary(?) Constraints on Computer Models of Natural Language Understanding

PHILIP A. GLOTZBACH

> The system-design approach that is common to AI and other styles of top-down psychology is beset by a variety of dangers, . . . [two of which are]:
> 1. mistaking *conditional* necessities of one's particular solution for completely general constraints (a trivial example would be proclaiming that brains use LISP; less trivial examples require careful elucidation).
> 2. restricting the performance of one's system to an artificially small part of the "natural" domain of that system and providing no efficient or plausible way for the system to be enlarged.
>
> Daniel C. Dennett, *Brainstorms*

INTRODUCTION

Technology mediates our lives in countless ways—most notoriously, on occasion, by coming between us and ourselves or our goals and so preventing or inhibiting some form of self-realization. To emphasize such worries is to represent one possible attitude toward or "way of being-with technology";[1] however, other possible relations between ourselves and the artificial world we create are available. More frequently, one hopes, technology mediates by enabling us to realize some purpose or goal that might otherwise be unattainable if not completely inconceivable. Technology also mediates our lives by helping to inform our self-image, stamping its indelible impression upon the stories we tell ourselves about ourselves. Consider tools. Weizenbaum argues that in addition to their more obvious functions tools serve as "pregnant symbols" that "symbolize the activities they enable," serve as models "for [their] own reproduction," and represent in concrete ways the skills required to use them. Weizenbaum concludes:

> The tool as symbol in all these respects thus transcends its role as a practical means toward certain ends: it is a constituent of man's *symbolic recreation of his world* [*sic*; emphasis added]. It must therefore inevitably enter into the imaginative calculus that constantly constructs his world. . . . [Tools] constitute a kind of language for the society that employs them, a language of social action.[2]

Extending this line of thought, Weizenbaum (and, more recently, Bolter) argues that the remarkable tool that is the computer has attained a special status as symbol, metaphor, and generator of new language.

Computers are unique not just because they are "autonomous machines"[3]—a characteristic they share with clocks, thermostatically controlled furnaces, and so on—but because at the deepest level of analysis they are realizations of universal, abstract machines whose sole function is to process information. We immediately recognize the affinity of this function with our own psychological processes, abstractly described.[4] Furthermore, we often find ourselves forced to describe and predict the behavior of a computer program by taking up what Dennett terms the "intentional" stance.[5] In so doing, we may ascribe to the program knowledge or beliefs about the world, goals and intentions, rational strategies for realizing those goals on the basis of available knowledge, and so on. For example, by relying on such ascriptions one can combat a chess-playing computer, rather as one would a human opponent—that is, by trying to identify its preferences for certain strategies over others, by finding gambits to which it is blind, and by recognizing and attempting to block its offensive plans. None of this forces one to think that such an artificial system "really" has beliefs in the same way that we do. The point is that the intentional stance makes available powerful explanatory resources that enable one to predict and even explain the behavior of a complex system more successfully than would otherwise be possible using less powerful lower-level representations of how the system behaves. However, because the intentional stance is precisely the one we adopt when dealing with persons, it comes as no surprise that, as Bolter puts it, the computer is constantly serving as a metaphor for the human mind or brain: psychologists speak of the input and output, sometimes even the hardware and software of the brain; linguists treat language as if it were a programming code; and everyone speaks of making computers "think."[6]

In the emerging discipline of cognitive science, a field that deploys the resources of cognitive psychology, philosophy, linguistics, computer science, and neuroscience in attempting to produce an adequate account of mental functioning, we find even more explicit and self-conscious reliance upon computational models of mind. Specifically, work in artificial intelligence (AI) has spawned computational accounts of problem solving, natural language understanding, perception (usually vision), knowledge representation, memory, and other psychological processes that have been taken up even by cognitive theorists with scant interest in trying to construct the actual systems that are the focus of AI research proper. There are many reasons behind the current popularity of such theories,[7] and I shall have a bit more to say about this. But first, I want to concentrate upon a specific difficulty I find in the attempts to understand language in this emerging research tradition. To articulate this difficulty it will be necessary to consider, in addition to issues in the theory of language itself, related matters in perception theory and epistemology.

UNDERSTANDING LANGUAGE/KNOWING THE WORLD

In a seminal paper in linguistics and the philosophy of language, Katz and Fodor conjecture that for any "piece" of real world knowledge it is possible to imagine some utterance *S* which that knowledge serves to disambiguate. More broadly, this means that any instance of real world knowledge *K* may be needed to interpret some sentence *S*, and conversely, that *K* will be a necessary part of the knowledge that a speaker must have in order to understand *S*. Katz and Fodor conclude that "a semantic theory cannot distinguish in principle between a speaker's knowledge of his [sic] language and his [sic] knowledge of the world."[8]

This thesis is almost universally accepted both by contemporary philosophers of language and AI researchers concerned with "natural language processing." For example, one of the most notable early successes in this area, Winograd's SHRDLU program, explicitly embraces Katz and Fodor's principle by utilizing a detailed representation structure to define objects and relations constituting a "blocks world." The program handles discourse about a world containing only "toy objects like blocks and balls" of different sizes and colors, a table upon which the objects are arranged, and a robot arm capable of manipulating the objects.[9] The resulting data base serves as an integral part of a program aimed at "understanding" natural language sentences about the blocks world domain. Here the concept of *understanding* functions operationally: the program is taken to "understand" when it correctly rearranges the blocks on command or responds appropriately, in colloquial English, to a question about the past or present states of the blocks world.

Let us consider in more detail the information SHRDLU has available. It represents the constituents of the blocks world in terms of predicate assignments to a set of basic objects $(B_1, B_2, \ldots,_n)$, which specify their color, size, shape, and so forth. Location in the blocks world is represented using a "three-dimensional coordinate system with coordinates ranging from 2 to 1200 in all three dimensions." The result is a "symbolic description, abstracting those aspects of the world which are relevant to the operations used in working with it and discussing it."[10] In effect, the program treats the statements constituting the model as axioms in a deductive system and manipulates blocks world objects by *deriving* appropriate results (comparable to deriving theorems) in response to commands and questions. This process is facilitated by populating the microworld only with objects definable precisely and completely in geometrical terms. A good deal of the "knowledge" embodied in the program, however, resides *not* in the individual statements of the world model, but in the operative *procedures* that interconnect the concepts with one another and contain information necessary to carry out actions.

Schank, Schank and Abelson, and others[11] have advanced the attempt to construct a knowledge-driven natural language understanding system by developing programs capable of interpreting sentences that require knowledge of relatively *less* structured contexts than Winograd's geometrical blocks world. The resulting programs (attempt to) represent such knowledge of the world in two ways. First, they employ "conceptual dependency" rules for mapping natural language sentences onto a representation scheme intended to capture their semantic content

in terms that are *interlingual* [emphasis added] and as neutral as possible. That is, we will not be concerned with whether it is possible to say something in a given language, but only with finding, once something is said, a representation that will account for the meaning of that utterance in an unambiguous way and one that can be transformed back into that utterance or back into any other utterances that have the same meaning.[12]

The resulting semantic representations are constructed out of (what Schank believes to be) concepts basic to the understanding of any natural language. Such concepts include a limited set of primitive ACTS such as ATRANS, for "Abstract TRANsfer of possession," PROPEL which relates to the application of force to objects, and various others.[13] These basic concepts should not be confused with the natural language verbs that they sometimes resemble. Instead, they are intended to represent the *underlying meaning* of any action or event describable in natural language. It is one of the interesting corollaries of Schank's program that comparatively few primitive acts, in conjunction with sometimes quite complex auxiliary descriptions, are held to accomplish the semantic work performed in natural languages by extensive classes of verbs.

One of Schank's major insights concerns the way in which people go about understanding sentences. He argues that as one listens to or reads a sentence one generates a set of specific expectations concerning what is likely to occur next. Such expectations frequently have a *syntactic* component—for example, sometimes one reaches a point in a sentence where one expects the next word to be a verb; for the most part, however, such expectations pertain to the *meaning* of what is being heard or read. By imposing substantial constraints upon interpretation, such semantic expectations help to account for much of the efficiency of our natural language understanding. Thus, when we encounter a potentially ambiguous sentence, our semantic expectations generally bring one interpretation to the fore, such that we typically fail even to notice that other interpretations might be possible. Settling quickly upon one option saves us the trouble of generating all possible interpretations and then choosing among them.

Schank represents the semantic content of a sentence by determining which of its conceptual elements depend upon which. Once identified, such dependency relations enable the system both to predict what semantic information might be coming next and, more important, permit it to narrow the choice of interpretation in a principled way. Schank identifies a number of syntactic relations that have this function—for example, active verbs tend to take objects, or some objects denoted by noun phrases can perform actions. But once again, the most important dependency relations turn out to be *semantic*. Thus consider the following sentence pairs:

1a. Tully took the book to New Haven.
1b. Tully took the train to New Haven.

2a. After he had insulted him, Burt gave Ernie a black sweater.
2b. After he had insulted him, Burt gave Ernie a black eye.

Though they differ only slightly on the surface, to understand these sentences one must assign them quite distinct semantic interpretations. We can do this, in part, because we know the linguistic conventions giving different connotations to "took" in the verb phrases "took the book" and "took the train." But we are able to *employ* such conventions successfully only because we can locate trains in the class of vehicles in which one can *take a ride*, while placing books in the class of things that one might *take with one* on a trip. Similarly, real world knowledge of the differences between black sweaters and black eyes is needed to disambiguate the possible meanings of "gave" as it occurs in 2a and 2b. Moreover, competent speakers of English would infer from 2a that Burt wanted to make amends, while concluding from 2b that adding injury to insult signals continuing hostility. All this constitutes one more application of Katz and Fodor's original contention: even when trying to write a set of rules mapping natural language sentences onto the highly abstract conceptual dependency structures, the distinction between linguistic and nonlinguistic knowledge goes by the board.

Second, Schank and Abelson employ the concept of *scripts* to extend the conceptual dependency approach to contexts involving patterns of ritualized or formalized activity. This second way of representing knowledge of the world requires one to write the rules governing behavior in, for example, a restaurant—including various loops and branches to handle different options (accepting the meal or returning it to the chef; eating the food in the restaurant versus taking it out; leaving a tip or not). Scripts detail sequences of related actions that together constitute a familiar project or activity. Thus, they represent the ways in which events in a larger sequence are connected and establish a background against which odd or unexpected occurrences stand out.[14] Thus, if a story says that Jill went to a restaurant, the restaurant script will enable the inference that she probably was hungry; if the story says that she ordered and paid the check, the script will enable the inference that she probably ate her meal. If we are told that she ordered a live lobster, the script should be able to supply the information that it was cooked before she ate it (thus requiring information about differences between preparing lobsters and, say, clams on the half-shell); conversely, the program should recognize that Jill's ordering a live chicken would be anomalous.

CONSTRAINTS ON ARTIFICIAL REPRESENTATIONALISM

Despite differences in the three examples of natural language processing systems considered above—Winograd's SHRDLU, Schank's conceptual dependency analysis, and Schank and Abelson's scripts, they share an important feature. Each relies on the *explicit, a priori* specification of *all* relevant details of the microworlds with which they are concerned. Whether these details are represented in terms of operational procedures, conceptual dependency parsing rules, or scripts does not really matter. Critics of AI, most notably Dreyfus,[15] have focused considerable attention on this, claiming that such preselection of data eliminates a major portion of the epistemic labor that *human* natural language users must perform on their own, and so considerably reduces any achievements of the resulting programs. I want to explore this point by introducing the concept of an epistemically closed system.

An *epistemically closed system* is a context in which it becomes possible in principle to anticipate all the questions that might arise in discourse about that system and to insure that the necessary answers to those questions can be derived or otherwise discovered on the basis of information available in advance. An AI microworld is an epistemically closed system. That is not at all to say that constructing a data base specifying a microword or writing a program capable of processing natural language utterances about such a domain of discourse becomes trivial *in practice*. Formidable difficulties remain because even when an effective decision procedure for a problem exists in principle—for example, the problem of winning at chess is soluble (in principle) by exhaustive search—it cannot necessarily be employed in practice.[16] Hence the need for *heuristic* procedures capable of cutting down the search space to manageable size. The notable results that have been achieved in developing such procedures testify to the ingenuity, insight, and persistence of AI researchers.

Nevertheless, solving the problem of natural language processing relative to a microworld, understood *qua* epistemically closed system, is not *qualitatively* different from solving the problem of chess. First, both enterprises require solutions to *bounded* problems governed by rules of relatively *unbounded* strength. That is to say, even if they do not constitute in practice an effective procedure for winning, the rules of chess do represent a usable effective procedure for distinguishing legal from illegal moves. In natural language processing, the task of assigning interpretations to character strings about a microworld more closely approximates the problem of determining legality in chess than it does the problem of winning. Second, the more a priori restrictions one can place on a given domain of discourse, the more assurance one has of being able to develop rules that approach an *effective procedure*. Such restrictions, first, limit the scope of the problems a natural language processing system will have to answer and, second, assure that all necessary information can be made available in advance.

In Winograd's case, since both the blocks world objects and their relational properties are completely defined by their mathematical specifications, all relevant information about those objects is either provided explicitly or can be derived from the axioms plus current state descriptions. This makes it possible, for example, to employ a rule that marks false any statement not definable within the system. This stipulation works for the blocks world, but the real world equivalent would be an epistemic principle requiring one to regard any knowledge claim that could not be verified as false. (In effect, Descartes claims to follow such a principle in the *Meditations*; whether he actually does so is an open question.) In the case of Schank and Schank and Abelson, similar effects are achieved by choosing microworlds in which plausible constitutive rules can be spelled out in advance. Subsequent shortcomings are handled by modifying those rules as necessary in the development phase. The result is a system able to handle any questions that the authors think might reasonably be asked about the microworld.

Persons operate within epistemic limits as well. However, when we interpret natural language we produce solutions to problems of potentially *unbounded* complexity using computational and informational resources that are in some ways quite limited relative to those of contemporary computers. Consider, for example, the uncertainties facing anthropologists who go to study a people whose culture is technologically centuries removed from their own and whose language they (initially) have no idea how to speak. The situation is only somewhat less problematic when readers first encounter a new form of literature, such as Joyce's *Finnegans Wake*. Even a Sunday newspaper presents an imposing range of epistemic demands. Understanding a typical series of articles might require a familiarity with professional sports, contemporary politics, space technology, film, international monetary policy, and religious fundamentalism. Furthermore, and perhaps most important, there is no way to specify in advance which questions might *reasonably* be asked of the reader of those articles. The point is that the information processing problems indicated here are *qualitatively* different from those investigated by researchers working with microworlds; consequently, they require solutions qualitatively different from those currently available in AI.

These remarks might appear to overlook the fact that, because AI natural language processing research is still in its infancy, researchers have pursued appropriately modest goals. But let me emphasize that the real question here concerns not the complexity (or lack of it) of the domains of discourse presently under study, but rather the *extendability* of the solutions produced in such research to increasingly difficult and interesting contexts. For example, can any current research program be extended by networking systems designed to deal with different microworlds? This might indeed be possible, but it raises the further difficulty of deciding which system to invoke to deal with a particular fragment of discourse. Developing an artificial system capable of solving this problem rquires a solution to the so-called setting selection problem, the problem of how speakers decide upon or interpret the *context* of a specific episode of discourse. As far as I know, no one has even the most remote idea of how to construct such a solution.[17]

The microworld frameworks themselves generate additional problems. A language processing system designed to handle a microworld represents a relatively specific solution to a specifically posed problem; generally such systems break down when unanticipated questions are encountered. This difficulty results not just from limitations of microworld representational schemata; it also follows from the fact that present natural language processing systems are designed to answer only certain types of questions about their microworlds—as indicated previously, just the questions that their designers consider "reasonable." For example, a program dealing with major league baseball would typically handle queries about batting averages and so on, while having nothing to say about the aesthetics of different team uniforms or ballparks. How many additional microworld-*cum*-language-processing-program systems would be required to simulate the baseball-relevant linguistic competence of a ten-year-old American child? How many would be required to simulate the latter's overall linguistic competence? How many programmer years would be needed to write the rules for these systems? We must remember that a primary stated purpose of the AI natural language processing research program is to gain insight into *human* linguistic abilities. So far, one finds little to suggest that the AI successes in dealing with epistemically closed systems represent any real progress toward the larger goal.[18]

A more promising proposal for extending current results comes from Schank who suggests that we must someday allow natural language processing systems to learn about the world on their own. This would be more efficient than extending their "knowledge" by modifying existing programs from the outside. Important objections have been raised against the claim that present natural language processing systems know *anything* about the world at all.[19] But let us leave this question aside for now and concentrate upon the issue of learning itself.

LEARNING, PERCEIVING, AND COGNITIVE SCIENCE

It goes without saying that some form of perceptual linkage to the world will be required to approximate the kind of learning that grounds human linguistic competence. By contrast to the a priori propositional specification of AI microworlds, we enjoy access to a practically infinite source of a posteriori information about the macroworld we inhabit. I believe that we begin to understand some of the significant difficulties afflicting research on AI language understanding by exploring several key contrasts between these two quite different ways of obtaining knowledge about a "world."

Three features of our perceptual relation to the world pertain to the matter at hand. First, the information potentially available to us is *quantitatively inexhaustible*. We always have the option of obtaining more of the same (kind of) information we already have. There are practical reasons why we cut short the search, but these have to do with constraints on time, memory, interests, computational resources, and so on—not constraints upon the information actually available. Second, and even more important, the information available to us is *qualitatively inexhaustible*. That is, we always have the option of asking *new* or previously unimagined questions about the world, questions that our perceptual abilities and other epistemic resources permit us to answer. This is quite different from current artificial systems, which are effectively restricted to handling only those questions relevant to information already contained in the representational structures specifying their microworlds. Third, the information available to us through perception always effectively outruns our ability to represent it *explicitly*—either to ourselves, by noticing available details, or to others through description. This in-principle unrestricted access to information about the world makes possible the linguistic competence that even young children typically display.

These three aspects of human perceptual competence constitute an ordered set of tasks that are increasingly difficult for designers of artificial systems to model. In principle, the first problem could be solved by giving an artificial system access to additional sources of environmental input or information. It is hardly trivial to say precisely how this is to be done, and particularly, to determine which principles of selection are to be employed. But the project is similar in key ways to ongoing attempts to enlarge and link data bases. The second goal of *qualitative* enhancement, the ability to obtain *new kinds* of information, could be achieved by developing a general solution to the problem of category learning or, more modestly, by devising ways to relax the rigid category structures employed by current data bases. Although Minsky's frames proposal[20] and Schank and Abelson's work on scripts (the latter grows out of the former) represent steps toward this goal, we have yet to see a solution that provides the required degree of flexibility. But the third contrast indicated above constitutes AI's most difficult challenge. This is the fact that our ability to obtain perceptual information about the environment systematically outruns our capacity to *represent* that information in propositional form, as opposed to artificial systems that characteristically rely upon propositional representation structures in order to extract the information in the first place.

To see what is at stake here, consider a typical human situation involving vision. Imagine yourself running on a path through a woods in late winter. The trees are quite bare of leaves, and so you can see for some distance in any direction. There are intermittent patches of ice on the asphalt path, but for the most part the surface is clear. Despite the fact that you are wearing gloves, it is cold enough to make your hands uncomfortable. You are running at a relaxed pace and generally thinking about other matters. How are we to describe your perceptual awareness? You are monitoring your body: the condition of your hands (to be sure that your fingers don't become too cold), your sore left knee (a little pain is all right, more would mean problems), the changing ratio of fatigue to the sense of physical well-being, and so on. Surely you are also paying attention to where you are going. You avoid the odd fallen branch, pothole, or patch of ice, and you notice persons coming the other direction long before they approach. At the same time, you are also aware of your less immediate surroundings: occasionally you see birds, perhaps a deer several hundred yards distant, as well as other items of interest. Yet you are not actively searching for anything out there. You are primarily thinking about, say, the paper you will try to finish when the run is over.

Notice that the preceding description is partial at best. It would take little effort to catalog additional objects of perceptual awareness, but it is not at all clear how the description could ever be completed—or, indeed, what that might even mean. Second, one is not exactly aware of all these things at once; there is something like a sequential cycling of attention from one object of interest to the next. At the same time, however, there is a kind of simultaneity to the experience: one avoids the ice patch while wondering why one's left hand feels colder than the right, and at the same moment one's attention is drawn instantly to the owl taking flight twenty yards off to the south. Third, and most important here, one could not begin to put all of this into words as it is occurring (or even afterward, for that matter). Any region or item in the perceptual background is a candidate for attention, but the sheer volume of information available as one moves through the environment prevents one's dealing explicitly with everything of which one is somehow perceptually aware. In short, unless we are prepared a priori to preclude the possibility of perceptual awareness where the object is not explicitly articulated, visual perception must establish a fundamental level of *pre*conceptual and *non*propositional[21] relations to one's environment and body. This primordial awareness of the environment supports an indeterminate number of possible conceptual or propositional "acts" at any moment, but it does not itself entail or require any one of them. Any attempt to analyze or simulate visual perception not addressing the issue of such preconceptual and/or nonpropositional awareness risks failing "to distinguish sharply between experiences with eyes opened and eyes closed."[22] More to the point, I suggest that the flexibility provided by our nonpropositional access to the perceived world enables us to deal with it as an epistemically *open* system. If this is correct, then our impressive linguistic competence might be explainable only in relation to this prelinguistic, preconceptual perceptual relationship to our environment. Consequently, it will be necessary to remove the a priori constraints imposed by the representational assumptions characteristic of standard AI research programs if genuine progress in natural language understanding is to be achieved.

What are current prospects that AI research into machine perception might enable artificial systems to gain anything like the kind of access to the world I have just described? Does one find any general awareness of the issues raised here? If one examines mainstream AI work on vision with these questions in mind, one finds little reason for optimism.[23] Instead, one finds in AI approaches to perception precisely the kind of restrictions identified previously in the discussion of natural language processing. That is, one encounters programs designed to recognize specific and usually quite limited sets of characteristics of the scenes presented as input. (Most typically, the input is two-dimensional display.) Target characteristics are invariably identified in advance, and propositional representation frameworks are the result. Some programs begin with specific descriptions of possible geometrical objects and attempt to identify such objects when they occur in input displays. Other programs begin, not with specific representations of objects, but with line-finding and vertex-labeling routines. These have proved to be remarkably powerful in distinguishing different objects in a cluttered display, but it has been difficult to make them recognize nonrectilineal shapes. Still other programs have exploited the shadows cast by objects as sources of information about three-dimensional objects. But in all such cases the situation remains as indicated earlier. Some such programs have been able to solve pattern-recognition problems common in industrial and robotics applications. Yet their pragmatic success in such contexts should not obscure the fact that they yield as output precisely the kind of propositional, preselected representations that I have contended are *inadequate* to ground an understanding of human natural language "processing."[24]

We continue to encounter the same fundamental restrictions in part because mainstream AI researchers have embraced constraints imposed upon them by existing programming languages and standard *serial* processing computer hardware.[25] As a consequence, those psychological theories of language and perception that have emerged from AI, or that have been most strongly influenced by this research, have fallen prey to Dennett's two dangers presented in the opening quotation of this paper. That is, any AI theorists are guilty of:

(1) mistaking *conditional* necessities of [their] particular solution for completely general constraints.
(2) restricting the performance of [their] system[s] to . . . artificially small part[s] of the "natural" domain[s] of [those] system[s] and providing no effective or plausible way for the system[s] to be enlarged.

But rather than belabor this point with respect to existing AI research, let me attempt to place these results in a broader conceptual context.

AI researchers have not concentrated so intently upon propositional, computational representation systems just to accommodate serial processing computer architecture, and existing programming languages. (After all, the most popular languages such as LISP and PLANNER were invented specifically for AI applications.) Nor have cognitive scientists outside of AI proper chosen to regard mental processes as computational procedures ranging over mental representations simply out of respect for the undeniable accomplishments of AI.[26] Rather, *both* groups have received substantial aid and encouragement from the representationalist tradition that has dominated psychology and the philosophy of mind at least since the seventeenth century and, according to some, since the time of Plato. This tradition has consistently looked upon mental events and states as particulars existing entirely in the mind. Frequently (it is hoped), mental particulars correspond with particulars states and events external to the mind, but even so the mind remains an autonomous realm with its own laws that must be studied as though it were divorced from physical reality.[27]

Despite its numerous philosophical problems, representationalism continues to dominate psychological[28] as well as philosophical theorizing, thus encouraging researchers to ignore the nonpropositional and preconceptual side of perception, largely because of the absence of credible, let alone successful, alternatives to the received view. For example, twentieth-century philosophy produced the phenomenological movement that attempted to ground epistemology and the philosophy of mind in descriptions of "lived experience." Unfortunately, even a phenomenologist such as Merleau-Ponty, whose work was informed throughout by a detailed knowledge of the psychological literature of his day, failed to show how his alternative to traditional views could be reconciled with reasonable scientific demands for experimental support and linkage to other levels of analysis (e.g., the neurophysiological level).[29] In psychology, traditional mentalistic theorizing retreated under the withering fire of the behaviorists who dominated American psychology for a good part of the current century. And yet mentalism has now reemerged as the dominant conceptual approach, largely because of behaviorism's clear inability to account for everything that psychology takes it upon itself to explain. And the dominant contemporary mentalist models remain representational ones.

So, what are we to conclude? I believe that three points emerge from this discussion:

1. Nothing I have said undercuts Katz and Fodor's fundamental insight that we cannot distinguish, even in principle, between a speaker's knowledge of a language and his or her knowledge of a world. Furthermore, we have every reason to think that this thesis applies with equal force both to humans and to artificial language understanding systems.

2. To design an artificial natural language-understanding system capable of performing beyond the bounds of microworld-based programs, it will be necessary to provide it with knowledge of the (macro)world that is not tied to a rigid, propositional representational scheme. Ideally, such an epistemic structure would be fundamentally *relational*—that is, it would be marked by schemata (for want of a better term) connecting it to the world that it purports to be knowledge *of*. The closer such connection schemata approximate the kind of *causal* connections that link our own perceptual experience to the world the better.

3. The world that the artificial system tries to understand must be treated as an epistemically *open* system.

There would be a reason to believe that these requirements could be fulfilled, first, only if we can identify a theory of perception capable of showing how biological systems achieve the indicated perceptual results and, second, only if we can envision a way to model that theory in an artificial system. In the concluding section of this paper, I shall suggest one direction in which an AI researcher might look for help in attempting to accommodate these conclusions.

PERCEPTION AND LANGUAGE WITHOUT REPRESENTATION

Over the past several decades many psychologists and philosophers have become interested in the so-called ecological approach to perception developed by the psychologist James J. Gibson.[30] I believe that the ecological position represents both a viable alternative to traditional theories of perception as well as a potential source of suggestive new ideas for AI.

Gibson's approach to perception differs radically from traditional representational theories. First of all, he emphasizes the need for an *evolutionary* account of an animal's perceptual abilities. That is, Gibson insists that we begin theorizing about perception by investigating the ways in which perceptual skills enhance an animal's fitness in relation to the environment in which its species evolved.

Consequently, the ecological approach focuses upon an animal's ability to pick up environmental information pertinent both to its special needs and to the risks associated with its ecological niche. Gibson thus emphasizes the *discriminative* function of perception: he characterizes as the primary function of perception the ability to discriminate among different objects, conditions, and states of affairs in the distal environment in terms of their *relational* qualities that are salient to the animal's requirements.[31]

Second, in stressing perceiver-environment relations, Gibson denies any role in his theory to the concept of *internal representation*. Perception, he argues, cannot be understood in terms of relations between a perceiver and some mediating internal representation scheme or perceptual hypothesis system; rather, perception is said to establish a *direct* relation between perceivers and their environment. In seeking to construct a viable account of direct perception, Gibson in effect rejects the propositional bias of the received view. For the propositional content in traditional theories enters in precisely at the point where the perceiver is supposed to construct an internal representation scheme of the external environment. In arguing for the concept of direct perception, Gibson thus tries to account for the nonpropositional aspects of our experience emphasized in the previous section.

Familiar problems plague any attempt to construct a theory of direct perception, and the Gibsonian program suffers no shortage of critics.[32] However, let us set the objections aside by noting that even if the problems associated with the concept of direct perception were to prove tractable, it still would remain to be seen whether the Gibsonian theory could be modeled in an artificial system. If this could be done, it might offer a way around the narrow representational constraints which I previously identified as hindering a solution to the problem of natural language understanding. What would have to be done to accomplish this?

First, implementing the Gibsonian theory in an artificial system would require more than modeling those aspects of perception with which AI workers have primarily been concerned, namely, its intellectual or conceptual aspects. Instead, some way would have to be found to simulate the role played by the perceiver's *body* in normal perception. For as Merleau-Ponty has stressed, perceivers first establish perceptual contact with the world through their bodies—specifically, in terms of the practical possibilities for action that the interaction between body and world makes available. Succeeding at this level of physical interaction with the world (a level that, one presumes, "lower" animals never transcend) requires, once again, just the kind of nonpropositional experience that I have highlighted. Second, this bodily access presents the animal with a perceived world composed of *relational* as opposed to absolute properties. For example, the animal sees an opening large enough for it to enter, not an opening of a certain objective size that is larger than the objective size of its body; it feels a branch that is strong enough to bear its weight; it walks cautiously on a surface on which it might easily slip; some objects in the world are edible, others inedible; some are possible mates, others are sexual rivals; and so on.[33] The point is that such determinations seem not to depend upon *prior* characterizations of the objective properties of the objects in question. Psychologists working to develop Gibson's ecological program are generating an increasing number of empirical studies aimed at discovering just how perceivers accomplish such perceptual tasks without relying upon intermediate representational structures. AI researchers could find this growing literature suggestive if they are prepared to approach the subject of perception with a new set of questions.

What actual implications for system design, if any, follow from the present discussion? Would an artificial system intended to incorporate this alternative approach to perception have to be mobile and capable of interacting with a physical environment on the basis of hard-wired "needs" and "vulnerabilities" supplied by the system designer? Not necessarily, perhaps. A good first step would be a conceptual exploration of the possibilities of nonpropositional, "fuzzy" representation schemes that have already attracted the notice of some AI researchers. But if one abandons the claim that an element of a knowledge system must be stored either as specific proposition, procedure, or "meta-fact,"[34] what is left? The answer to this question is not at all clear. Perhaps it is only essential that the artificial system be capable of modeling the detection of relational environmental properties through something like a causal link to the outside. However, the concept of a *relational* property becomes empty in the absence of a specific body—the perceiver's body—to serve as a concrete reference point. And it is not at all clear how this is to be realized in the context of AI.

Another problem lurking in the background of this discussion— or, possibly just another way of articulating the need to include a relational component in perception—concerns the *intentional* characteristics of biological perception and the difficulty of reproducing this in artificial systems. Minimally, the internal states of such a system should be referred to in the external environment by the system itself (as opposed to having this done by the operator who reads the system's output), and there is no reason to believe that this can ever be accomplished merely by adding additional representational elements to an internal representational structure. If a system already begins with nonreferring data structures, then augmenting those structures with additional (nonreferring) representations is not likely to change the system's basic nonintentional character. I suggest, instead, that the question of intentionality needs to be explored by experimenting with different kinds of causal links between the system and the world. Perhaps one of the principal advantages of the connection machine architectures presently being investigated is that they force us both to rethink the nature of the machine-world interface and to reconsider the nature of internal representations and programming.[35]

I write under no illusions. There is precious little guidance here, and readers would not be blamed for feeling more frustrated than enlightened. But the present discussion will have served its purpose by articulating a set of questions about the linkage between natural language understanding and perception and, perhaps, by providing some reasons why those questions should be taken seriously. I have maintained that the representational approach to perception is a dead end and that genuine progress will be made only by those willing to explore alternative paths. If I am correct, then success with natural language processing programs also ultimately depends upon a willingness to address the kind of issues adumbrated here and, specifically, a willingness to reconsider longstanding representational preconceptions that may be obstructing future progress. From this perspective, system designers interested in considering a change in direction can be well served even by a preliminary analysis of such issues.

This essay began with a few words about the extent to which dominant computational metaphors and the resulting conceptual models have influenced our understanding of ourselves as thinking beings. Interestingly enough, it concludes with the suggestion that attempts to create artificial models of perception and language might benefit from a new look at how biological systems deal with perception. One can only hope that this kind of dialectical movement will increasingly characterize our attempts to understand both ourselves and the possibilities inherent in our computing systems.[36]

NOTES

1. See C. Mitcham, "Three Ways of Being-With Technology," pp. 31–59 above.

2. Joseph Weizenbaum, *Computer Power and Human Reason* (San Francisco: W. H. Freeman and Co., 1978), pp. 18, 36. See Martin Heidegger, *Being and Time*, trans. John Macquarrie and Edward Robinson (New York: Harper and Row, 1962), for an extended treatment of this theme. J. D. Bolter also discusses the place of the computer and the metaphors of its grounds in contemporary self-understanding in *Turing's Man: Western Culture in the Computer Age* (Chapel Hill: University of North Carolina Press, 1984).

3. Weizenbaum, *Computer Power and Human Reason*, p. 24.

4. See Jerry Fodor, "The Mind-Body Problem," *Scientific American* 244 (January 1981): 1.

5. Daniel C. Dennett, "Artificial Intelligence as Philosophy and Psychology," *Brainstorms* (Cambridge: MIT Press, 1981), pp. 6–8.

6. Bolter, *Turing's Man*, p. 11.

7. See, e.g., Dennett, "Artificial Intelligence," pp. 109–26; Fodor, "Mind-Body Problem"; and J. Haugeland, ed. *Mind Design: Philosophy, Psychology, Artificial Intelligence* (Montgomery, V.: Bradford Books, 1981).

8. J. J. Katz and Jerry Fodor, "The Structure of a Semantic Theory," *Language* 39, no. 2, pt. 1 (April–June 1963): 179.

9. Terry Winograd, *Understanding Natural Language* (New York: Academic Press, 1972).

10. Ibid., p. 118

11. Terry Winograd, "A Procedural Model of Language Understanding," in *Computer Models of Thought and Language*, ed. R. Schank and K. Colby (San Francisco: W. H. Freeman and Co., 1973), p. 167.

12. Roger Schank, "Conceptualizations Underlying Natural Language," in *Computer Models of Thought and Language*, pp. 187–247; Roger Schank, *The Cognitive Computer: On Language, Learning, and Artificial Intelligence* (Reading, Mass.: Addison Wesley Publishing Co., 1984); and Roger Schank and R. Abelson, *Scripts, Plans, Goals and Understanding* (Hillsdale, N. J.: Lawrence Erlbaum Associates, 1977).

13. Schank, "Conceptualizations Underlying Natural Language," p. 191.

14. Schank, *The Cognitive Computer*, pp. 98, 103. Schank's claim that his conceptual dependency representations are "interlingual"—i.e., are not tied to any specific language—raises a number of theoretical issues that go to the heart of his enterprise. However, these and related questions about the meaningfulness of the components of his representation scheme fall outside the scope of the present inquiry.

15. Ibid., p. 114.

16. See Hubert L. Dreyfus, *What Computers Can't Do: A Critique of Artificial Reason*, rev. ed. (New York: Harper, Colophon Books, 1979).

17. See Herbert Simon, *The Sciences of the Artificial*, 2d ed. (Cambridge: MIT Press, 1981). In the chess case, the exponential growth of the search space quickly overwhelms any conceivable computational resources.

18. Indeed, Katz and Fodor contend that there can be no *general* formal solution to this problem.

19. See, e.g., Schank, "Conceptualizations Underlying Natural Language," for a statement of purpose. Dreyfus likens the microworld analyses to climbing a tree and thinking that one has made a "first step" toward reaching the moon.

20. Cf. Schank, *The Cognitive Computer*; and, e.g., Dreyfus, *What Computers Can't Do*, and John R. Searle, "Minds, Brains, and Programs," *The Behavioral and Brain Sciences* 3, no. 3 (1980).

21. Marvin Minsky, "A Framework for Representing Knowledge," in *The Psychology of Computer Vision*, ed. P. Winston (New York: McGraw-Hill, 1975), pp. 211–80.

22. I do not wish to suggest that the concepts of "preconceptual" and "nonpropositional" awareness are identical, but for present purposes the distinction between them can be ignored.

23. W. Reitman, R. Nady, and B. Wilcox, "Machine Perception: What Makes It So Hard for Computers to See?" in *Perception and Cognition: Issues in the Foundations of Psychology*, Minnesota Studies in the Philosophy of Science, 9, ed. C.W. Savage (Minneapolis: University of Minnesota Press, 1978), pp. 65–87.

24. See A. Barr and E. A. Feigenbaum, *The Handbook of Artificial Intelligence* (Los Altos, Calif.: William Kaufmann, 1981) vol. 3; M. Boden, *Artificial Intelligence and Natural Man* (New York: Basic Books, 1977); and Winston, *The Psychology of Computer Vision*, for surveys of the literature. Reitman et al. represent an interesting alternative to general trends in AI. It becomes less surprising that researchers in AI have been insensitive to this issue when one realizes that even philosophers have made few attempts to attack the question of perception from this revisionary perspective. Maurice Merleau-Ponty, *The Phenomenology of Perception*, trans. Colin Smith (New York: Humanities Press, 1962), and Ludwig Wittgenstein, *Philosophical Investigations*, trans. G. E. M. Anscombe (Oxford: Basil Blackwell, 1953), represent significant exceptions. Mark Johnson, *The Body in the Mind: The Bodily Basis of Imagination and Language* (Chicago: University of Chicago Press, 1987) presents the most detailed and compelling argument to date for the thesis expressed here.

25. For an intriguing break with tradition, which upon first inspection appears at least potentially compatible with the thrust of the present analysis, see T. Winograd, and F. Flores, *Understanding Computers and Cognition: A New Foundation for Design* (Norwood, N. J.: Ablex Publishing Corp., 1986), p. 44.

26. See Lawrence Davis, "From Artificial to Natural Intelligence: A Philosophical Critique," pp. 196–208 below, for an indication of how radically different forms of computer architecture might permit a different approach to some of these problems.

27. See, e.g. Fodor, "The Mind-Body Problem"; and Jerry Fodor and Z. Pylyshyn, "How Direct is Visual Perception?: Some Reflections on Gibson's 'Ecological Approach,'" *Cognition* 9 (1981): 139–96.

28. Dreyfus and Heidegger argue the historical point. Jerry Fodor, "Methodological Solipsism Considered as a Research Strategy in Cognitive Psychology," in *Representations* (Cambridge: MIT Press, 1981), approves of this methodology, whereas Searle, "Minds, Brains, and Programs," does not.

29. See, e.g., L. Kaufmann, *Sight and Mind: An Introduction to Visual Perception* (Oxford: Oxford University Press, 1974), chap. 1.

30. As has already been suggested, I believe that it is possible to show how this can be done and, hence, to provide a role for phenomenological accounts such as that provided by Merleau-Ponty in the psychological analysis of perception. For a first approximation, see Philip A. Glotzbach and Harry Heft, "Ecological and Phenomenological Contributions to the Psychology of Perception," *NOUS* 16, no. 1 (1982): 108-21.

31. See J. J. Gibson, *The Senses Considered as Perceptual Systems* (Boston: Houghton Mifflin, 1966); and J. J. Gibson, *The Ecological Approach to Visual Perception* (Boston: Houghton Mifflin, 1979).

32. In this aspect, Gibson's view relates most closely to earlier *behavioral* approaches to perception.

33. See, e.g., Fodor and Pylyshyn, "How Direct is Visual Perception?"

34. See Gibson, *The Ecological Approach to Visual Perception*, chap. 8; and virtually all of Merleau-Ponty, *The Phenomenology of Perception*.

35. Minsky, "A Framework for Representing Knowledge."

36. See Davis, "From Artificial to Natural Intelligence: A Philosophical Critique"; see also Reitman et al., "Machine Perception: What Makes it so Hard for Computers to See?"

From Artificial to Natural Intelligence: A Philosophical Critique

LAWRENCE DAVIS

INTRODUCTION

This essay surveys and analyzes two important ways that artificial intelligence researchers have represented human knowledge. The points to be presented are based on four years' experience as an artificial intelligence researcher, but much of my training before then was in the field of philosophy, which has influenced the form of the survey and critique. In particular, I have organized the survey to support my belief that the evolution of our knowledge representation techniques is moving us closer and closer to a true theory of human thought, a theory that most philosophers past and present have thought false.

FROM LOGIC CIRCUITS TO EXPERT SYSTEMS

One of the most astounding things about computers is their basic simplicity. Computers are built from transistors, and transistors are extremely simple logic circuits. When one programs a computer at the levels that are quite close to the physical makeup of the machine, one comes to appreciate their simplicity in a very real way. The things one can do there involve manipulating numbers in low bases—base 2 or base 8. One may store such numbers at various locations, move them around, add them, increment them, decrement them, or create lists of such operations, with the option to jump from one place in the list to another depending on the result of numerical tests. When a person studies programming at this level, the first exercise might consist of writing a routine that multiplies two numbers or one that reads sequences of numbers from one place and stores them in another.

At the lower levels of computer programming, this sort of exercise can be quite challenging. One worries about what is located where, how to reference it, when to perform operations, and what happens when different parts of one's program reference and alter the contents of the same location. Curiously, at these lower levels, other worries are almost nonexistent. Since this is the basic level at which we tell computers what to do, it is clear why people have said that computers are really very dumb machines. "They aren't smart, because we have to tell them *everything*." Yet what we tell them is usually wrong. It is difficult to translate our ideas into low-level programs, and difficult to find out what has gone wrong when we observe the effects of the inevitable mistranslation.

Low-level computer languages are not natural languages for us. In order to make it easier for us to tell computers what to do, most programmers use higher-level languages. A high-level language is one with a syntax of its own, which is translated into a lower language by a program called an interpreter.

There are hundreds of higher-order computer languages. FORTRAN, COBOL, and BASIC are by no means the best of them, but they are among the most widely used, and they share a number of important properties. With them, it is relatively easy to program a computer to carry out numerical computations and vary its behavior based on tests of the data it has been given. Because of them, we no longer think seriously of calculating logarithmic tables or the trajectories of bodies in flight ourselves. In fact, many schoolchildren can write programs that direct a computer to carry out, in a few seconds, work that consumed years of human labor in the past and yielded less accurate results.

With the use of higher-order languages, we have made computers our betters in the area of mathematical calculation. More recently, computer scientists have developed high-level languages that are modeled on modern logic, and these languages have allowed computers to carry out reasoning processes that are very much like some human reasoning processes.

The media have paid a good deal of attention to these reasoning processes, publicizing the fact that computer programs called expert systems have found mineral deposits, configured computer installations, and diagnosed diseases better than many of the humans who currently do those things professionally. What is impressive about these programs is not that they are built up painstakingly from elementary logical operations, for they are not—not, at least at the level on which we deal with them. Expert systems contain sentence-like representations of knowledge that, operating together, produce behavior like that of the experts whose expertise the knowledge represents.

I will produce a simple expert system here to demonstrate the way such programs work. The program itself, possibly the first moral expert system ever written, will be composed of rules designed to determine which of two actions is better to perform.

1. X is better than Y if
   X violates a strict subset of the moral constraints that Y violates.
2. X is better than Y if
   X violates moral constraints that are less serious than the moral constraints that Y violates.
3. X is better than Y if
   X violates no moral constraints, and
   X increases the world's good more than Y.
4. Robbery is less serious than murder.
5. Lying is less serious than murder.

We can run this program by loading it into an expert system driver and, when given the prompting message ASK ME SOMETHING, type in, "Is sending my sister a postcard from Colorado Springs better than sending her a letter bomb from Colorado Springs?" The expert system driver will look at rule 1 and see that it needs to find out about the moral constraints that each action violates. I have not told it how to do that in the rules above, so it will ask me. Suppose I tell it that sending the postcard violates no constraints, but sending the letter bomb violates a constraint against committing murder. The system will immediately be able to tell me that the answer to my question is YES, sending the postcard is better.

The system will also be able to tell me whether it is better to steal the postage for the postcard or to murder to get the postage. The first rule will not be relevant, but the second rule will trigger use of the fourth rule, and the system will be able to advise me, without asking me anything at all. The system will be able to tell me whether stealing the postage for the postcard is better than lying to get it, if I am able to say which action greater increases the world's good. In fact there are a great many decisions that this system will be able to advise me on, as long as I can give it the information it needs.

One thing that makes such a system interesting, and what makes it different from writing a program in BASIC or FORTRAN, say, is that I can add and delete information from the system and it will still run. Whatever errors it has will be errors in results, not errors in operation. Programming an expert system involves adding a sentence to a list of sentences, taking a sentence out, or editing a sentence that it already has. The result will always be computer-executable. This is quite a different experience from programming in other ways.

Another source of interest in expert systems stems from a recurrent theme in Western thought. A tremendous amount of attention has been paid to the development and theory of logical reasoning in the West. The discovery that many of the complicated theorems of geometry that had been found through empirical observation could be derived by applying simple rules of reasoning to concepts characterized by simple axioms was one of the most influential insights in the West. Since that time, most mathematicians have learned their fields by studying the basic axioms and definitions, then working through proofs based on them to conclusions, even though the advanced conclusions, not the axioms and definitions, are the subject matter of interest.

It is a recurrent and widely agreed-upon theme in Western philosophy that the preferred technique for presenting and developing one's theories is logical. In fact, the analytical tradition, one of the two major streams of contemporary Western philosophy, has been characterized by some of its members as nothing but the attempt to analyze our concepts and the way we hook them together with logic.

This picture of us as thinking properly only insofar as we are rational beings employing well-understood concepts together with logic to draw out their consequences has led us in the last few decades to create logic-based computer languages. LISP, the current language of choice for American artificial intelligence users and the language most American expert systems are based on, is a programmable version of the lambda calculus. PROLOG, the preferred language for many European and Japanese artificial intelligence researchers, is an executable, restricted version of the predicate calculus, the logic system learned by students in most contemporary philosophy classes.

Representing formal human knowledge in these languages, as opposed to mathematically oriented languages like FORTRAN, is a very natural thing—philosophers have been doing it for decades. If logical thinking is the highest form of thought, then logic programming languages appear to be the culmination of millennia of philosophical investigation, and expert systems written in those languages would seem capable of doing the things we do when we are at our best—acting as rational entities. No wonder the media have given expert systems a good deal of attention.

This concludes our discussion of knowledge representation through logic-based systems, from the extremely simple sort of logics operating on binary numbers to the logics of predicates used in expert system development. I have attempted here to convey the sense that many researchers have had, of working to represent human thought in the best way possible. Given our received dictum that logic is the preferred mode of thought, given the results of the sciences in which formal representation and formal reasoning have played a major role, and given the promise of contemporary expert systems to equal and surpass us as users of the knowledge we have gained, such researchers are certainly well supported in believing that theirs is the best way to explore the nature of human thinking.

DIFFICULTIES WITH LOGICAL MODELS OF THOUGHT

Before considering other ways of representing knowledge, it is important to draw a distinction between a system's capabilities and its natural use. Two computers running different languages of the sort we have been discussing are equivalent in a sense formalized by the mathematician Alan Turing: the actions of a computer programmed in one can always be simulated by a computer programmed in the other. An interpreter translates the activities of higher-level languages into lower-level ones, and it is a simple matter to translate in reverse. What is important for our purposes, however, is the natural use to which we would put such languages. It is extremely difficult to perform higher-level tasks in a low-level language. It is difficult to carry out reasoning in the predicate calculus using FORTRAN. What computer scientists are after when they attempt to model human intelligence are computer architectures or languages that make computations like those that humans perform happen naturally.

It is the view of some computer science and cognitive science researchers that logic-based models are not the most natural ones for representing the processes that underlie human thought. Let us consider four of their arguments for this conclusion.

1. Our own abilities do not seem to spring from elementary logic circuits of the sort that our computers are built from. As our understanding of the brain develops, the actions of the neurons in it

appear to involve less and less passing of input signals through and-gates and or-gates to arrive at output ports labeled "jump" or "don't jump." Rather, it seems that neurons are connected to hundreds of other neurons, and their firing is a partly random process, rather than a precise response to logical inputs. The activity of responding to external stimuli seems to consist in finding some linked pattern of neurons that can emerge from random firings between neurons excited by the inputs, rather than getting an input signal to follow one logical path through the neural circuitry of the brain to its conclusion. It seems that our thoughts may emerge in the way a pattern emerges out of chaos, or a statue emerges from a sculptor's marble block.

2. If this model of our thinking processes is correct, then our employment of logical thinking patterns as our preferred mode may not guarantee the quality of the results. There is some empirical evidence that this is indeed the case. The history of mathematics is littered with proofs of theorems presented by experienced mathematicians that later turned out to be flawed. The four-color theorem of topology, for example, was "proved" by the mathematician Kempe in 1879. This "proof" stood for eleven years, until Heawood found an error in it. The number of purported proofs of this and other classical theorems of mathematics is great, as is the number of correct proofs that were not accepted for years by the mathematical community.

The point here is that our best-trained reasoners are not always able to reason correctly when employing logic, nor are they always able to assess the correctness of reasoning when examining instances of it. However beneficial its use may be, we do not seem to have an infallible criterion for telling when we are using it correctly. It may be a good medium for the transmission of results in the field, but it is by no means clear that the results are arrived at with the use of logic.

3. A part of the attraction of logic has been that when one reasons with the rules of logic from true premises, one always derives true conclusions. In many of our daily operations, however, we appear to use uncertain and conflicting principles in the pursuit of ill-defined goals. The logics that have been employed in expert systems force us to ignore a great deal of the contingency that obtains in these situations. We are able to attach probabilities to the conditions and conclusions of our expert system rules, but we are not easily able to represent the contingent nature of much of our probabilistic knowledge. When we reason with probabilities, we depart from the part of certain knowledge, and many of the attractions of the logical approach dim when this departure is made.

4. The most creative aspects of intelligence do not appear to lie in

deduction, but in coming up with the premises our deductions will use. Arthur Conan Doyle provides an interesting illustration of this point, for in the Sherlock Holmes stories he continually confuses them. Holmes is made to attribute his successes to deduction, when in fact they are most frequently due to his ability to generate *some* hypothesis, no matter how far-fetched, that accounts for all the clues. Holmes uses deduction to test the hypotheses for consistency with the clues, but his leaps of insight were creative, not deductive. It was an easy matter for Holmes to explain his true deductions— reasoning from trouser stains to a client's itinerary, for example— and we agree with Holmes that such reasoning is elementary once we hear how it was done. His solution to mysteries like that of the speckled band is quite another matter.[1] How such solutions arise in human brains has not been understood by us. We have no expert systems to produce them, nor do we have a formal theory of them that we can build into our machines. We are blind to their arising in a way that we are not blind to our trains of logical thought, and yet so much of our intelligence is composed of these mysterious upwell- ings of inspiration that they, too, call out for computer counterparts if we are to have computer systems that learn and hypothesize creatively, as we do.

These considerations have inspired computer science researchers and cognitive science researchers to consider other ways of repre- senting knowledge. In what follows, I discuss two of them.

CONNECTION NETWORKS

One new way of modeling thought takes the notion of a network of connections as its metaphor, together with simplifying assump- tions about the ways things in those networks will be connected to many other units. The connections between units may be variably weighted, and the units may be in the ON state or the OFF state. Such connection networks are used in various ways. A common approach, however, is to set some of the units in the ON state and some in the OFF state as the network begins operation. At random intervals, each unit in the network sums the connection strengths of the units it is connected to. If the units that are on have greater strength than the units that are off, the unit in question turns itself on. Otherwise, it turns itself off. A chance factor is involved in the network's operation, because the order in which units decide what state to be in is a random one. In some systems, the decision whether to turn on or off has a chance component added as well. Because of such chance factors, given the same initial state it is possible for a connection network to end up in quite different final states, depending on the order in which units compute. Connection networks can have a number of stable states, and the stable state that results when a network starts with a set of units set to be ON is the result of computing, given those units as input.

Which stable state is likely to result for a given initial setting of the system depends on the connections and weights of connections between units. The stable states themselves have been compared with concepts, in that a number of similar input settings can produce an identical stable state, which characterizes or categorizes those input settings.

The process of arriving at a stable state given an initial setting of ON units is of interest because it seems like certain behaviors that we ourselves undergo. The initial ON and OFF settings could be produced by some stimulus from the environment. The weights on connections could have been derived by concept learning. The final configuration of units that are on can be considered to represent a concept of some sort. The way in which individual units turn themselves on and off until a stable configuration is reached is like our groping processing of ambiguous input until we arrive at a conclusion about it—deciding we have seen a black cat in the dark rather than a dog, on the basis of a slight movement and a noise, for example.

This example illustrates an important feature of connection networks: the initial stimulus given them need not be very complete or accurate for the system to work correctly. A brief glimpse of a cat in the dark can be enough for us to decide that a cat was there. Like a connection network, we tend to fill in the rest of the features from those we have grasped. What is important in our programming, like that of the connection network, is the amount of linkage that certain features have had before in the presence of cats. A logic-based system would be required to state the subset relationships between catlike characteristics and other characteristics in a great deal of detail in order to acquire this sort of behavior. A connection network gets it for free, as a result of its probabilistic operation.

Another interesting feature of the connection network is that units in themselves do not represent concepts. A given unit can be a member of hundreds of final states; what counts is the whole set of units that end up in the ON position. Logic-based systems tend to have single variables or parameters that hold this sort of information. In a traditional, data-base style of knowledge representation, our information about an entity might include a CAT bit that is on or off. If that bit were to be lost or changed the system would no longer know whether the entity was a cat. The loss of a unit hardly degrades the performance of a connection network. Because concepts are distributed over networks of units that are highly connected, single units can be lost and the resulting network will continue to respond pretty much as it did before the unit was lost. This feature of connection networks corresponds nicely with the findings of some neurophysicists that certain of our mental functions are degraded in proportion to the amount of our cerebral matter that is removed.

The connectionist model of mental operation is quite provocative. It allows us to create systems that will classify a wide variety of contradictory, erroneous, and incomplete data with the aid of a probabilistic series of internal firings settling down finally in a state from which the classification can be read. This sort of thing is very difficult to do in a logic-based system.

On the other hand, a connection network is very difficult to program. We cannot easily understand the meaning of any connection's weight, since a given connection can be part of a number of "concepts." We are also cannot easily explain a connection network's operation, occurring as it does through randomly generated firings. Before the creation of computers, determining what such networks do when they are functioning would have been very difficult, because they are composed of so many units with so many interactions.

One hope for the research along these lines is that as commercially valuable results are derived from it, it will be necessary for us to write programs that generate human-comprehensible summaries of the systems' operation. The problem of deriving such summaries may shed some light on the problem we have of representing our own neural interactions to ourselves. Our brains do something in order to represent their activities to us in language. When we ourselves are faced with the problem of representing connectionist processes comprehensibly, we may learn some surprising things about how it is done in the brain.

The connectionist approach forces us to think about logic in a different way. Suppose we think of ourselves as having connection networks that encode our learning from previous experience as weights on connections between units in our brains. How would the process of carrying out logical inference be implemented in such a system? One way would be to implement a logical rule like *modus ponens*, for example, as the excitation of a group of units in the presence of premises of the form $P \rightarrow Q$ and P that result in the excitation of units corresponding to Q. The probabilistic nature of the network stabilization process would not guarantee Q every time, but we could approach 100 percent certainty as closely as we liked by setting the system parameters appropriately. As the concepts bound up in the P and Q were less well understood, or as they appeared in propositions deeply embedded in the system of weights and connections, the system would tend to make mistakes in the processing with greater and greater probability. Such facts might form explanations for the failure of mathematicians to detect errors in their reasoning—they believe the conclusion so deeply that their system will tend to it with or without logic. Similarly, experts who believe in a conclusion at odds with the result of a logical proof might have a good deal of trouble believing in the steps it involves, because many of the steps would create interference with the belief structure they already possess.

GENETIC ALGORITHMS

The connectionist model may illuminate the way we think, given an encoding of our experience, but the problem of encoding experience remains. The last system I want to discuss models learning from experience in a provocative way. The technique is that of using genetic algorithm to carry out automated learning. Genetic algorithm, modeled on the biological processes of evolution, were developed by John Holland and others.[3] Struck by the way that from a multitude of designs of terrestrial creatures nature "learned" to produce good ones relatively quickly, Holland has been applying nature's techniques to other sorts of problems.[4]

Genetic algorithms operate by creating a population of individuals that represent solutions to the problem to be solved. Each individual's "fitness" is assessed by evaluating how well it does in the environment of the problem. A new population is created by reproducing members of the old, with the fittest tending to reproduce often and the least fit tending not to reproduce. During the reproductive process, genetic operators are randomly applied to the parents. These operators include random mutation, reorderings, and combinations of genetic matter from two different parents. When such systems are well engineered, the populations tend to evolve better and better individuals, and they do so quite rapidly, because genetic matter independently evolved can be combined when two parents produce a single offspring.

Holland and his collaborators have applied the techniques of genetic algorithms to the problem of learning from an environment in systems they call "classifier systems." Classifier systems are made up of a number of rules, each consisting of a precondition and an action. The assumption is that the rules form part of a system functioning in an environment. Over time, it will learn to improve its performance by discovering the environment's regularities and exploiting them.

Information from the environment and information generated by the rules themselves are combined on a message list. The medium of representation used by the researchers thus far has been quite simple: messages are simply strings of zeros and ones—bit strings. Each rule's precondition is a template that it attempts to match against a bit string on the message list. If its precondition matches some message or other, the rule posts its own message during the next cycle of operation. From time to time, the classifier system is rewarded or punished by the environment, and the reward or punishment is distributed among those rules that were active at the time the response occurred. Periodically, genetic operators are applied to the rules in the classifier system, so that rules with poor performance will be replaced by rules exploring new responses, or varying responses that the system has already learned.

What happens when a classifier system runs is very interesting. An initial collection of randomly generated rules will have some rules that perform better than average in the environment, some that do not, and some that set the stage for the better-performing rules to operate. The better-performing rules tend to survive and reproduce, replacing the ones that do badly. When rules are combined, their offspring acquire combinations of bit patterns that have been found useful by their parents. Chains form between rules that detect interesting states of the environment and rules linked to behaviors that are useful in the presence of such states. These chains are reinforced if the behaviours are appropriate in the environment. As a result, classifier systems tend to form complicated patterns of behavior that persist in situations they have learned to handle, because new rules cannot compete with the stengths of successful parents. Genetic operators lead the systems to try out responses in unfamiliar situations, so that the systems tend to generalize and extend their areas of competence.

The rules in a classifier system function quite differently in this respect from the rules in the expert systems discussed earlier. In a logic-based expert system, a single rule was triggered by a query; it triggered a single rule, or a request of the system user. At every step of the program's running, only one rule was employed, and it led to a single conclusion. The flow of logic was rigidly controlled by the program that oversaw the expert system's operation. If there was a bad rule in the set—a rule that concludes the wrong thing, for example—its conclusion would be used by all the other rules in the system if it was the first one the system tried to establish as its conclusion. In such a case, even the rules that worked correctly would never be tried. Using a single-path approach to expert systems, a single bad rule could ruin the performance of the entire system.

In classifier systems, all the rules operate simultaneously. Rules that are doing the correct thing and rules that are not will all have a chance to post their messages, and over time the ones that work correctly will have increased in strength over others, and the incorrect ones will tend to be replaced. Classifier systems can easily accommodate inconsistent, incomplete, and confusing inputs in a way that logic-based expert systems cannot.

Classifier systems have been used to learn to monitor the flow of gas through a natural gas pipline, maximizing profit in the face of daily and seasonal variations in demand and in the face of random leaks. Given the rules of poker, one system learned to outperform a well-known logic-based program that had been a standard in the artificial intelligence field. Classifier systems are being used for experimental vision systems, and they are being experimented on as robotic control systems. Their proponents believe that classifier systems are the best computer models so far devised for modeling learning from an environment. Significantly, these results are derived from systems that are random in nature and alogical in operation, features that may help to explain the poor results philosophers have had in their attempts to produce logics of induction.

CONCLUSIONS

This is a very interesting time in which to be a philosopher and a computer scientist, for it is one of those rare times when the results of philosophical research and the perceived needs of society have coincided. A tremendous amount of theory about the nature of human thought is now being transferred from philosophy to the sciences, and in the reverse direction as well. Philosophers have often played the role in our society of investigating those features of society and the sciences that shed light on human nature. The development of computers, with their potential for modeling our capacities, is a phenomenon that I believe will stimulate a good deal of useful philosophical research. We have noted that philosophical research in logic and knowledge representation has had a great impact on the development of expert systems and higher-order computer languages. My prediction is that, as we develop systems that perform induction and determine what actions to perform through random processes rather than logical inference, and as the success of those systems in modeling human processes becomes evident, there will be attempts to rethink the nature of human rationality within the community of philosophers. I cannot say what the results of all these interactions will be, but I anticipate them with the keenest interest.

NOTES

1. See "The Adventure of the Speckled Band," by Sir Arthur Conan Doyle. In the denouement Holmes describes the process through which he solved the mystery. Phrases like ". . . My attention was speedily drawn . . . to this ventilator" and "The discovery that this was a dummy, and that the bed was clamped to the floor, instantly gave rise to the suspicion that the rope was there as a bridge" suggest intuition and creativity rather than deduction.

2. A good place to begin reading about connection networks is *Parallel Distributed Processing: Explorations in the Microstructure of Cognition*, ed. David Rummelhart and James L. McClelland (Cambridge: MIT Press, 1986).

3. John Holland, et al., *Induction: Processes of Inference, Learning and Discovery* (Cambridge: MIT Press, 1987). This text is the best source of information about genetic algorithms and classifier systems.

4. Lawrence Davis, ed. *Genetic Algorithms and Simulated Annealing*, Pitman Research Notes in Artificial Intelligence (New York: Wiley 1987), contains several papers commenting on the relationship between connection networks, genetic algorithms, and human cognition.

Select Bibliography

D'Alembert, Jean Le Rond. *Preliminary Discourse to the Encyclopedia of Diderot*. Translated by Richard N. Schwab and Walter Rex. Indianapolis, Ind.: Bobbs-Merrill, 1963.

Arendt, Hannah, *The Human Condition*. Chicago: University of Chicago Press, 1958.

Bacon, Francis. *New Organon*. Indianapolis, Ind.: Bobbs-Merrill, 1960.

———. *Selected Writings*. Edited by Hugh G. Dick. New York: Modern Library, 1955.

Benson, Ian, ed. *Intelligent Machinery Theory and Practice*. New York: Cambridge University Press, 1986.

Berdjajew, Nikolai. *Der Mensch in der technischen Zivilization*. Vienna: Amandus, 1948.

Berger, Peter, et al. *The Homeless Mind: Modernization and Consciousness*. New York: Random House, 1974.

Bergmann, Frithjof. "The Future of Work." *Praxis International* 3 (October 1983): 308–23.

Berland, K. J. H. "Bringing Philosophy Down from the Heavens: Socrates and the New Science." *Journal of the History of Ideas* 47, no. 2 (April–June 1986): 299–308.

Bolter, J. D. *Turing's Man: Western Culture in the Computer Age*. Chapel Hill: University of North Carolina Press, 1984.

Borgmann, Albert. *Technology and the Character of Contemporary Life: A Philosophical Inquiry*. Chicago: University of Chicago Press, 1984.

Borgmann, Albert, with the assistance of Carl Mitcham. "The Question of Heidegger and Technology: A Critical Review of the Literature." *Philosophy Today* 31, no. 2/4 (Summer 1987): 99–194.

Bugliarello, George, and Dean B. Doner, eds. *The History and Philosophy of Technology*. Introduction by Melvin Kranzberg. Urbana: University of Illinois Press, 1979.

Burke, Edmund. *A Philosophical Inquiry into the Origin of Our Ideas of the Sublime and Beautiful*. 2d ed. London: R. & J. Dodsley, 1759.

Byrne, Edmund F. "Robots and the Future of Work." In *The World of Work*, edited by Howard F. Didsbury, Jr. pp. 30–38. Bethesda, M.: World Future Society, 1983.

Cornford, F. M. "Greek Natural Philosophy and Modern Science." In *The Unwritten Philosophy*. New York: Cambridge University Press, 1967.

Davis, Lawrence, ed. *Genetic Algorithms and Simulated Annealing*. Pitman Research Notes in Artificial Intelligence. New York: Wiley, 1987.

Deleuze, Gilles. *Cinema I: L'Image mouvement*. Paris: Les Éditions Minuit, 1983.

Dennett, Daniel C. *Brainstorms*. Cambridge: MIT Press, 1981.

Derrida, Jacques. *La Carte postale: de Socrates à Freud et au-delà*. Paris: Flammarion, 1980.

Descartes, René. *Discourse on Method*. Translated by Laurence J. Lafleur. Indianapolis, Ind.: Bobbs-Merrill, 1950.

Dewey, John. *Experience and Nature*. New York: Dover Publications, 1958.

————. *The Quest for Certainty: A Study of the Relation between Knowledge and Action*. New York: G. P. Putnam's Sons, 1929.

Dickson, David. *The Politics of Alternative Technology*. New York: Universe, 1975.

Diderot, Denis. "Art." In *Encyclopedia: Selections*. Translated by Nelly S. Hoyt and Thomas Cassirer. Indianapolis, Ind.: Bobbs-Merrill, 1965.

Dreyfus, Herbert L. *What Computers Can't Do: A Critique of Artificial Reason*. Revised edition. New York: Colophon Books, 1979.

Ellul, Jacques. *The Technological Society*. Translated by John Wilkinson. New York: Alfred D. Knopf, 1964.

Feibleman, James K. *Technology and Reality*. The Hague: Martinus Nijhoff, 1982.

Ferkiss, Victor C. *Technological Man: Myth and Reality*. New York: Mentor, 1969.

Ferré, Frederick. *Philosophy of Technology*. Englewood Cliffs, N.J.: Prentice Hall, 1988.

Fischer, Frank, and Carmen Sirianni, eds. *Critical Studies in Organization Theory and Bureaucracy*. Philadelphia: Temple University Press, 1984.

Florman, Samuel. *Blaming Technology: The Irrational Search for Scapegoats*. New York: St. Martin's Press, 1981.

Fodor, Jerry, and John Haugeland, eds. *Mind Design: Philosophy, Psychology, Artificial Intelligence*. Cambridge: MIT Press, 1981.

Friedmann, Georges. *The Anatomy of Work*. Translated by Wyatt Rawson. New York: Free Press, 1964.

Gibson, J. J. *The Ecological Approach to Visual Perception*. Boston: Houghton Mifflin, 1979.

————. *The Senses Considered as Perceptual Systems*. Boston: Houghton Mifflin, 1966.

Glotzbach, Philip A., and Harry Heft. "Ecological and Phenomenological Contributions to the Psychology of Perception." *NOUS* 16, no. 1 (1982): 108–21.

Grene, Marjorie. "Hobbes and the Modern Mind." In *Philosophy in and out of Europe*, pp. 155–66. Berkeley: University of California Press, 1976.

Gorz, Andre. *Paths to Paradise: On the Liberation from Work*. Translated by M. Imrie. London and Sydney: Pluto Press, 1985.

————. "The Tyranny of the Factory: Today and Tomorrow." In *The Division of Labor*, edited by Andre Gorz, pp. 55–61. Sussex: Harvester, 1976.

Hanks, Joyce Main. *Jacques Ellul: A Comprehensive Bibliography*. Greenwich, Conn.: JAI Press, 1984.

Hayim, Gila J. *The Existential Sociology of Jean-Paul Sartre*. Amherst: University of Massachusetts Press, 1980.

Heidegger, Martin. *Being and Time*. Translated by John Macquarrie and Edward Robinson. New York: Harper and Row, 1962.

———. *The Question Concerning Technology and Other Essays*. Translated by William Lovitt. New York: Harper and Row, 1977.

———. "The Thing." In *Poetry, Language, Thought*, translated by Albert Hofstadter, pp. 165–86. New York: Harper and Row, 1971.

Heilbroner, Robert L. *An Inquiry into the Human Pospect*. 2d edition. New York: W. W. Norton, 1980.

Hill, Christopher T., and James W. Utterback. *Technological Innovation for a Dynamic Economy*. New York: Pergamon Press, 1979.

Holland, John., et al. *Induction: Processes of Inference, Learning and Discovery*. Cambridge: MIT Press, 1987.

Horkheimer, Max. *The Eclipse of Reason*. New York: Seabury Press, 1974.

Hume, David. *Essays*. Oxford and London: Oxford University Press, 1963.

———. *A Treatise of Human Nature: Being an Attempt to Introduce the Experimental Method of Reasoning into Moral Subjects*. Edited by L. A. Selby-Bigge. Oxford: The Clarendon Press, 1978.

Husserl, Edmund. *Cartesian Meditations*. Translated by Dorian Cairns. The Hague: Martinus Nijhoff, 1977.

———. *Ideas: General Introduction to Pure Phenomenology*. Translated by W. R. Boyce Gibson. New York: Collier-Macmillan, 1962.

Ihde, Don. *Existential Techniques*. Albany: State University of New York Press, 1983.

Illich, Ivan, and Barry Sanders. *ABC: The Alphabetization of the Popular Mind*. San Francisco: North Point Press, 1988.

Jacoby, Henry. *The Bureaucratization of the World*. Berkeley: University of California Press, 1973.

Jaspers, Karl. *Man in the Modern Age*. Translated by Eden Paul and Cedar Paul. Garden City, N.Y.: Doubleday, 1957.

Jennings, Humphrey. *Pandaemonium: The Coming of the Machine as Seen by Contemporary Observers, 1660–1886*. Edited by Mary Lou Jennings and Charles Madge. New York: Free Press, 1985.

Jonas, Hans. *The Imperative of Responsibility: In Search of an Ethics for the Technological Age*. Chicago: University of Chicago Press, 1984.

———. *The Phenomenon of Life: Toward a Philosophical Biology*. New York: Dell, 1966.

———. *Philosophical Essays: From Ancient Creed to Technological Man*. Englewood Cliffs, N.J.: Prentice Hall, 1974.

———. "The Scientific and Technological Revolutions: Their History and Meaning." *Philosophy Today* 15 (Summer 1971): 76–101.

Kant, Immanuel. *On History*. Translated by Lewis White Beck. Indianapolis, Ind.: Bobbs-Merrill, 1963.

Kohanski, Alexander Sissel, *Philosophy and Technology: Toward a New Orientation in Modern Thinking*. New York: Philosophical Library, 1977.

Kostas, Axelos. *Alienation, Praxis, and Techné in the Thought of Karl Marx*. Translated by Ronald Bruzina. Austin: University of Texas Press, 1976.

Kranzberg, Melvin, and Carroll W. Pursell, Jr. "The Importance of Technology in

Human Affairs." In *Technology in Western Civilization: The Emergence of Modern Industrial Society, Earliest Times to 1900*, 1: 3–11. Oxford: Oxford University Press, 1967.

Krohn, Wolfgang, Edwin Layton, Jr., and Peter Weingart, eds. *The Dynamics of Science and Technology: Social Values, Technical Norms, and Scientific Criteria in the Development of Knowledge*. Dordrecht: D. Reidel Publishing, 1978.

LeGoff, Jacques. *Time, Work, & Culture in the Middle Ages*. Translated by A. Goldhammer. Chicago: University of Chicago Press, 1982.

Lobkowicz, Nicholas. *Theory and Practice: History of a Concept from Aristotle to Marx*. Notre Dame, Ind.: University of Notre Dame Press, 1967.

MacIntyre, Alisdair. "Social Science Methodology as the Ideology of Bureaucratic Authority." In *Through the Looking Glass*, edited by Maria J. Falco, pp. 42–58. Washington, D.C.: University Press of America, 1979.

Mayo, Elton. *The Social Problems of an Industrial Civilization*. London: Routledge and Kegan Paul, 1949.

McDermott, John. "Technology: The Opiate of the Intellectuals." *New York Review of Books*, 31 July 1969, 25–34.

Manuel, Frank E., and Fritzie P. Manuel. *Utopian Thought in the Western World*. Cambridge: Harvard University Press, 1979.

Marcuse, Herbert. *One-Dimensional Man*. Boston: Beacon Press, 1964.

Marx, Karl, and Frederick Engels. *Collected Works*. Translated by Richard Dixon and others. New York: International Publishers, 1975–.

Marx, Leo. *The Machine and the Garden: Technology and the Pastoral Ideal in America*. New York: Oxford University Press, 1964.

———. "Reflections on the Neo-Romantic Critique of Science." *Daedalus* 107, no. 2. (Spring 1978): 61–74.

Merleau-Ponty, Maurice. *The Phenomenology of Perception*. Translated by Colin Smith. New York: Humanities Press, 1962.

de La Mettrie, Julien Offray. *Man a Machine*. La Salle, Ill. Open Court Press, 1912.

Michie, Donald, and Rory Johnston. *The Knowledge Machine: Artificial Intelligence and the Future of Man*. New York: William Morrow, 1985.

Michalos, Alex C., ed. *Philosophical Problems of Science and Technology*. Boston: Allyn and Unwin, 1974.

Minsky, Marvin. "A Framework for Representing Knowledge." In *The Psychology of Computer Vision*, edited by P. Winston, pp. 211–80. New York: McGraw-Hill, 1975.

Mitcham, Carl. "'The Love of Technology Is the Root of All Evil.'" *Epiphany Journal* 8, no. 1 (Fall 1987): 17–28.

———. "Philosophy and the History of Technology." In *The History and Philosophy of Technology*, edited by George Bugliarello and Dean B. Doner, pp. 163–201. Urbana: University of Illinois Press, 1979.

Mitcham, Carl, with the assistance of Jim Grote. "Current Bibliography in the Philosophy of Technology: 1975–76." In *Research in Philosophy and Technology*, vol. 4, edited by Paul T. Durbin, pp. 1–297. Greenwich, Conn.: JAI Press, 1981.

———. "Current Bibliography in the Philosophy of Technology: 1977–78." In *Re-*

search in Philosophy and Technology, vol. 6, edited by Paul T. Durbin. pp. 231–89. Greenwich, Conn. JAI Press, 1983.

Mitcham, Carl, and Alois Huning, eds. *Philosophy and Technology II: Information Technology and Computers in Theory and Practice*. Dordrecht: D. Reidel Publish, 1986.

Mitcham, Carl, and Robert Mackey, eds. *Bibliography of the Philosophy of Technology*. Chicago: University of Chicago Press, 1973.

Montesquieu, Charles. *The Spirit of the Laws*. Translated by Thomas Nugent. Edited by David Wallace Carrithers. Berkeley: University of California Press, 1977.

Nietzsche, Friedrich. *The Birth of Tragedy*. Translated by Walter Kaufmann. New York: Vintage Books, 1967.

———. *The Gay Science*. Translated by Walter Kaufmann. New York: Vintage Books, 1974.

Noble, David S. *Forces of Production: A Social History of Industrial Automation*. New York: Alfred Knopf, 1984.

Ormiston, Gayle L., and Raphael Sassower. "Language and Culture: The Context of STS Education." *Bulletin of Science, Technology & Society* 7 (1988): 754–57.

Opitz, Peter J. "Thomas Hobbes." In *Zwischen Revolution and Restauration*, edited by Eric Voegelin, pp. 47–81. Munich: List, 1968.

Ortega y Gassett, José. "Man the Technician." In *History as a System*, translated by Helene Weyl, pp. 87–164. New York: W. W. Norton, 1962.

Pacey, Arnold. *The Culture of Technology*. Cambridge: MIT Press, 1986.

Pitt, Joseph C. "The Epistemological Engine." *Philosophica* 32 (1983): 77–95.

———. "Galileo: Causation and the Use of Geometry." In *New Perspectives on Galileo*, edited by Robert E. Butts and Joseph C. Pitt, pp. 181–95. Dordrecht: D. Reidel, 1978.

Pope, Alexander. *An Essay on Man*. London: L. Gilliver, 1734.

Rousseau, Jean-Jacques. "A Discourse on the Arts and Sciences." In *The Social Contract and Discourses*, translated by G. D. H. Cole, pp. 125–55. New York: Dutton, 1950.

de Saint-Lambert, Charles François. "Luxury." In *Encyclopedia: Selections*, translated by Nelly S. Hoyt and Thomas Cassirer, pp. 203–32. Indianapolis, Ind.: Bobbs-Merrill, 1965.

Sartre, Jean-Paul. *Being and Nothingness*. Translated by Hazel Barnes. New York: Washington Square Press, 1968.

———. *Critique of Dialectical Reason*. Translated by Alan Sheridan-Smith. Edited by Jonathan Reé. London: New Left Books, 1976.

———. "Existentialism as a Humanism." In *Existentialism from Dostoyevsky to Sartre*, edited by Walter Kaufmann, pp. 354–68. New York: New American Library, 1975.

Schank, Roger. *The Cognitive Computer: On Language, Learning, and Artificial Intelligence*. Reading, Mass.: Addison Wesley, 1984.

Sibley, Mulford Q. *Nature and Civilization: Some Implications for Politics*. Itasca, Ill.: F. E. Peacock, 1977.

Simon, Herbert. *The Sciences of the Artificial*. 2d edition. Cambridge: MIT Press, 1981.

Singer, Charles, E. J. Holmyard, and A. R. Hall, eds. *A History of Technology*. 8 vols. Oxford: The Clarendon Press, 1954–78.

Snow, C. P. *The Two Cultures*. Cambridge: Cambridge University Press, 1964.

Sypher, Wylie. *Literature and Technology: The Alien Vision*. New York: Random House, 1968.

Trinkhous, C. E. *In Our Image and Likeness: Humanity and Divinity in Italian Humanist Thought*. 2 vols. London: Constable, 1970.

Veblen, Thorstein. *The Theory of the Leisure Class*. New York: Modern Library, 1934.

Voegelin, Eric. "Industrial Society in Search of Reason." In *World Technology and Human Destiny*, edited by Raymond Aron, pp. 38–48. Ann Arbor: University of Michigan Press, 1963.

———. *The New Science of Politics*. Chicago: University of Chicago Press, 1952.

Weber, Max. *The Protestant Ethic and the Spirit of Capitalism*. Translated by Talcott Parsons. New York: Scribners, 1958.

Weizenbaum, Joseph. *Computer Power and Human Reason*. San Francisco: W. H. Freeman and Co., 1978.

Whitehead, Alfred North. *Science and the Modern World*. New York: Mentor Books, 1948.

Winner, Langdon. *Autonomous Technology: Technics-Out-of-Control as a Theme in Political Thought*. Cambridge: MIT Press, 1978.

———. *The Whale and the Reactor: A Search for Limits in an Age of High Technology*. Chicago: University of Chicago Press, 1986.

Winograd, Terry. *Understanding Computers and Cognition: A New Foundation for Design*. Norwood, N. J.: Ablex Publishing, 1986.

———. "A Procedural Model of Language Understanding." In *Computer Models of Thought and Language*, edited by R. Schank and K. Colby, pp. 152–84. San Francisco: W. H. Freeman and Co., 1973.

———. *Understanding Natural Language*. New York: Academic Press, 1972.

Wittgenstein, Ludwig. *Philosophical Investigations*. Translated by G. E. M. Anscombe. New York: Macmillan, 1968.

Index

Abelson, R., 178, 180, 182, 185
Abstraction, powers of, 86
Agassi, Joseph, 15
Agriculture, 33, 34
AI. See Artificial intelligence
Aldiss, Brian, 93
Algorithms, genetic, 205
Alienation, 164, 166–72
Analysis: class, 142, 162; existential, 32, 55; metaphysical, 36, 39
Anámnesis, 109
Anaxagoras, 66
Andreae, Johann, 139
Anti-worker, 141
Aquinas, Thomas, 34, 42, 137
Archimedes, 38
Arendt, Hannah, 72 n.6
Aristotle, 34, 37, 38, 42, 49, 61–68, 82, 126, 137
Art, 45, 112
Artifact(s), 13, 14, 23, 37, 55, 70; alliance with artifice, 13, 16; alliance with habitat, 13, 16
Artifice, 13, 19, 21, 23, 44, 45, 52, 109, 112; alliance with habitat, 13, 16; technological, 18, 19, 20
Artificial intelligence (AI), 21, 177, 181, 183–91, 196, 199
Arts, 33; and enlightenment, 41; of luxury, 43; mechanical, 41
Astronomy, 124–28
Augustine, Saint, 44
Authenticity, technology and, 32
Authority, 156, 158, 170; rational-legal, 157, 165
Automation, 149
Autonomy, 176. *See also* technology: the autonomy of

Bacon, Francis, 17, 39–44, 49, 50, 62, 63, 66, 68, 95; as epistemological pragmatist, 43
"Bad faith," 165
BASIC, 198, 199
Baudelaire, Pierre Charles, 76
Being-with, 17, 31–59, 164
Being-with-others, 166
Bellers, John, 140
Bender, Frederic L., 20; on organization theory, 20; on technology and alienation, 20; on technology's relationship to bureaucracy, 20
Benedict, Saint, 136, 137, 151
Berger, Peter, 72 n.4
Bergerac, Cyrano de, 95
Bias, antilabor, 146
Biology, Aristotelian, 123
Bios theoretikos, 17, 65, 69, 71
Blake, William, 50, 51, 52
Blocks world, 178, 182
Bolter, J. D., 176
Braverman, Harry, 142, 143, 149, 162
Bureaucracy, 155–72; and the administration of power, 157; and communication, 166; critical theory of, 161–66; and power, 61; Weberian ideal of, 156
Burke, Edmund, 53
Burton, Robert, 139
Bush, Vannevar, 145
Butler, Samuel, 18, 97, 98
Byrne, Edmund F., 20; on the autonomy of technology, 20; on labor-saving devices, 20; on technology and responsibility, 20

Calvin, John, 136, 139
Calvinism, 139
Campenella, Tommaso, 95
Capitalism, 139, 140, 156, 157, 159, 162; bureaucratic, 170; and profits, 142
Cartesianism, 75, 76

Categories, 82
Causation, 66–68
Cenobites, 136
Chernobyl, 53
Childe, V. Gordon, 115 nn. 5 and 9
Christianity, 38–39, 40, 44, 52, 71;
    Baconian interpretation, 41; Enlight-
    enment version of Bacon's interpreta-
    tion, 41
Cinema, 108, 114
Class analysis, 42, 162
COBAL, 197
Cohen, I. B., 122
Coleridge, Samuel Taylor, 51, 80
Community, technology and, 15
Competition, 144
Computers, 176, 196, 197, 207
Concept(s): alliance with operations of
    technology, 113–14; alliance with
    sensible artifacts, 113; of technology
    and self, 13, 17
Consciousness, 88, 96; technical, 96;
    technological, 60–71, 69
Construction, 34; arts of, 37
Constructions, metaphorical, 87
Contemplation, the life of, 69
Content, semantic, 179
Context, 112
Cosmology, organic, 46
Creation, 112
Creativity, human, 150
Crowley, Ambrose, 140
Cultivation, arts of, 37
Culture, 84, 86

D'Alembert, Jean le Rond, 41, 49
Daniken, Erich von, 90–91
Darius, 84, 85
Davidson, Edward H., 79, 80
Davis, Lawrence, 21, 194 n.26, 195 n.36;
    on evolutionary theory of knowledge
    representation, 22; on genetic theory
    of knowledge representation, 22; on
    knowledge representation, 21
Deleuze, Gilles, 114
Democritus, 66
Dennett, Daniel C., 22, 175, 176, 187
Derrida, Jacques, 108, 116 n.13
Descartes, René, 17, 22, 62, 63, 66, 68,
    75, 81, 86, 182
Determinism: hard, 132; technological,
    132

Dewey, John, 112, 116 n.19
Dialectic, 170, 172
Dickens, Charles, 50
Diderot, Denis, 44, 95
Dilemma, prisoners', 146
Discourse, technology and, 14, 111
Dissimulation, 21, 112
Doyle, Arthur Conan, 202
Drake, S., 122, 124, 125
Dreyfus, Hubert, 181

Economic theories, neo-classical, 146
Ellul, Jacques, 17, 74, 74–101, 95, 96,
    117, 151 n.1
Engels, Frederick, 149
Engineering, genetic, 91–92
Enlightenment, 41–45, 48, 51, 52
Episteme, 109, 110
Epistemology, 177
Equipment, 19, 32; present-to-hand,
    32; ready-to-hand, 32
Ethic, work, 134, 136
Etymology, 104–13
Evans, Lawrence B., 144
Existential analysis, 32, 55
Existentialism, 163–72; as a social
    philosophy, 164
Experience, 103

Fabrication, 104; life of, 66, 69
Fact, science, 92
Factory Laws, 142
Familiar, the, 74; technological, 76
Fantasia, 18, 74, 75, 82, 85, 87, 93; and
    language and history, 82
Feeling, 46
Ferkiss, Victor C., 72 n.6
Fiction, science. See Science fiction
Fielder, Leslie, 92, 93
Florman, Samuel, 24 n.13
Flying saucers, 89–91
Fodor, Jerry, 22, 177, 180, 188
Fontenelle, Bernard de, 44
FORTRAN, 196, 197, 199
Framework, interpretive, 16
Francis, Saint, of Assisi, 137
Freedom, 165, 167, 169
Freud, Sigmund, 150
Fromm, Eric, 150, 153 n.45

Galileo, 123–29
Genetic engineering, 91–92

Genetic operators, 206
Geometry, 127, 128
Gernsback, Hugo, 92
Gesture, 85
Gibson, James J., 189, 190, 191
Glotzbach, Philip A., 21, 194n.30; on artificial intelligence, 22; on cognitive science, 22; on computational representation systems, 22; on ecological approach to AI, 22; on knowledge representation, 21; on perception, 22
Goldman, Paul, 163
Good, the, 64, 67
Gorz, Andre, 139, 149
Gourmount, Remy De, 80
Grene, Marjorie, 70
Group, 168
Group-in-fusion, 168, 169, 170

Habitat, 13, 14, 15, 19, 20; alliance with artifact(s), 13, 16; alliance with artifice, 13, 16; as structures of thought, action, and discourse, 14
Hall, A. R., 104
Hawthorne, Nathaniel, 79, 87
Heft, Harry, 194n.30
Hegel, G. W. F., 23, 88
Heidegger, Martin, 23, 31, 32, 39, 55, 72n.6, 193n.2; on techne, 116n.19
Heilbroner, Robert L., 71n.2
Herodotus, 85
Hill, Christopher T., 24n.13
Hiroshima, 53
Hobbes, Thomas, 45, 70, 71
Hoffman, Daniel, 80
Holland, John, 205, 206
Holmes, Sherlock, 202
Holmyard, E. J., 104
Homer, 35
Homme machine, l', 45
Homo faber, 45
Horkheimer, Max, 72n.6
"Human relations" movement, 158, 159, 160, 161
Hume, David, 24n.4, 42, 43
Husserl, Edmund, 18

Idanthyrsus, 85
Ideologies, 155
Ideology, organizational, 160
Ihde, Don, 115n.3
Images, 15, 75, 78, 85, 99, 108, 114

Imagination, 46, 51, 74, 80–89, 92, 95, 99; disciplined, 94; and divine insight, 98; as a faculty of apprehension in Kant, 81; as the ground of reason, 82–83; and Mystery, 82–101; and the transcendental, 93
Individual, 165; as a self-creative project, 166
Industrial Revolution, 46, 49, 50, 51, 135, 139, 141; and automation, 142, 145; introduction of the "Iron Man," 141–42
Ingenio, 86
Innovation, technological, 13
Institutions, 172; bureaucratic, 165, 168
Intelligence, artificial. *See* Artificial intelligence
Intersubjectivity, 166
Iterability, 21

Jefferson, Thomas, 34
Johnson, Samuel, 39
Jonas, Hans, 68
Jove, 83, 85, 88, 99
Joyce, James, 93, 182
Jung, C. G., 89, 90

Kafka, Franz, 93
Kant, Immanuel, 41, 80, 81; on Enlightenment, 41
Katz, J. J., 22, 177, 180, 188
Knowing-how, 110
Knowing-that, 110
Knowledge, 19, 43, 84, 110, 122, 123, 178, 180, 182; humanistic, 84; representational theories of (perception), 189; technical, 35; technological, 40
Kranzberg, Melvin, 115nn. 5–7
Kuhn, Thomas S., 126

Language, 84, 85, 177, 188; natural, 197; philosophy of, 177
LeGoff, Jacques, 151n.3, 152n.6
Lenin, V. I., 149
Life, the good, 61
Linguistics, 177
Lipperhey, Hans, 131n.9
LISP, 188, 199
Literacy, 95
Literature, transcendence of, 18
Lived-experience, 188

Logic, 92
Logos, 21, 32, 64, 111
Lovekind, David, 17; on imaginative literature, 18; on *la technique*, 17; on science fiction, 18; on Vico's conception of imagination, 18; on Vico's conception of reason, 18
Luther, Martin, 138

Machines, 148; autonomous, 20
McDermott, John, 119, 120
MacIntyre, Alasdair, 173 n.3
Management, 133, 134, 142, 161; scientific, 159, 161; "value neutral" science of, 158
Mann, Thomas, 93
Manufacture, 140
Marcuse, Herbert, 24 n.2, 72, 149, 150
Markovic, Mihalio, 155
Marx, Karl, 136, 139, 142, 143, 148, 149, 156
Marx, Leo, 58 n.48, 71 n.1
Marxism, 149, 167, 168, 169
Mayo, Elton, 156, 158, 160, 161
Meaning, will to, 163
Mechanics, Newtonian, 46
Memoria, 86
Memory, 85–86
Menken, H. L., 97
Mentalism, 188
Mentality, Cartesian, 76
Merleau-Ponty, Maurice, 22, 188, 190
Metaphor, 21, 23, 75, 78, 85, 99
Metaphysical analysis, 34, 39
Metaphysics, 45, 52, 64, 71
Method, Cartesian, 18, 75
Mettrie, Julien Offray de la, 45
Microworld frameworks, 183
Mill, John Stuart, 156
Milton, John, 50
Minsky, Marvin, 185
Mitcham, Carl, 17, 55 n.3, 192 n.1; on ancient skepticism, 17; on Renaissance and Enlightenment optimism, 17; on romantic ambiguity, 17
Mitsein, 164
Mobil Oil Company, 37
Model, connectionist, 202–5
Models, logic, 196
Montesquieu, Charles, 43
More, Thomas, 95

Morris, William, 150
Mumford, Lewis, 39
Mutus, 84
Mystérion, 17
Mystery, 80, 81, 99; technology and, 74–101
Mythos, 84
Myths, 33, 94

Narratives, 104, 113
Nature, 41, 43, 44, 45, 46, 52, 66, 67, 68, 69, 71; as a drama, 67; modern understanding of, 66
Newton, Isaac, 128
Nietzsche, Friedrich, 58 n.49, 103
Noble, David, 142
Nonpropositional relations, 186

Operators, genetic, 206
Optimism, Enlightenment, 53; Renaissance and Enlightenment, 33
Ora et labora, 136
Organization, 170
Ormiston, Gayle L., 17, 116 n.17; on art and techné, 18; on the issue of technology, 18; on technology and metaphor, 19; on translation, 18
Other, the, 80, 99
Otto, Rodolf, 81

Perception, 22, 177, 184, 185, 186, 189; and the body, 190, 191; direct, 190; ecological approach to, 189, 190, 191; nonpropositional, 188, 191; nonpropositional elements of, 22; preconceptual, 188, 190; and the relation to natural language understanding, 192
Philmus, Robert M., 95
Philosophy, moral, 49
Phusis, 38
Piercy, Marge, 150
Pitt, Joseph C., 19; on the autonomy of technology, 19
PLANNER, 188
Plato, 22, 23, 34, 35, 38, 44, 49, 61–68, 95, 109, 188
Plattes, Gabriel, 140
Plutarch, 38
Poe, Edgar Allan, 18, 76, 78, 80, 87, 88, 89, 93, 94, 97, 102; on the artist, the writer, 77; on the literary work, 78;

the paradigm of *la technique*, 77; and scientific religion, 79
Poetic act, 82
Poets, 85
Poiesis, 44; the distinction between divine and human, 44–45
Pope, Alexander, 44
Poster, Mark, 169
Power: knowledge and, 49; managerial, 156
Practices, technological structures and, 17
Praxis, 168, 169, 170
Preconceptual relations, 186
Presentation, 108
Pretext, 13
Principles, 14, 16, 69, 109, 110; of inquiry, 102; linguistic, 14; scientific, 14, 124; social and political, 14, 62
Production, mass, 140
Project, 163, 164, 168; possibilities of, 165
PROLOG, 199
Proudhon, Joseph, 139, 150
Proust, Marcel, 93
Providence, 86–87
Puritanism, 136, 139

Rabkin, Eric S., 92
Ragione, 18, 74, 93
Reality, 62, 75; material conditions of, 66; rational causes of, 66
Reason, 41, 50, 51, 70, 74, 75, 81, 82, 84, 89, 91, 95; life of, 69, 70; as technical, 96
Reciprocity, 168
Recollection, 109
Relations: nonpropositional, 186; preconceptual, 186; self-other, 167; syntactic, 179
Renaissance, 41, 45
Representation, 21, 206; of human knowledge, 21; knowledge, 177, 196, 199, 203
Representationalism, 175, 188, 191; artificial, 175, 181
Representations, semantic, 179
Reproduction, 21
Rilke, Ranier Maria, 39
Ritual, 86

Romanticism, 46, 50; and technology, 46
Rousseau, Jean-Jacques, 49, 50
Rule: interim, 135; ultimate, 135, 140
Rules, 206
Ruskin, John, 150
Ryle, Gilbert, 110, 116n.18

Sarabites, 136
Sartre, Jean-Paul, 20, 156, 164, 165, 166, 168, 169, 170, 172; existential Marxist critique, 172; existential Marxist theory of bureaucracy, 166–72; and mediated reciprocity, 20
Sassower, Raphael, 116n.17
Schaff, Adam, 153n.43
Schank, Roger, 178, 179, 180, 182, 184
Scholes, Robert, 92
Schumacher, E. F., 150
Science, 62, 63, 87, 109, 112, 123; cognitive, 21; natural, 34, 70; philosophy of, 122; the purpose of, 63–64; three types of science in Aristotle, 64–66
Science fiction, 76, 92, 93, 97
Sciences, arts and, 42, 84
Scientia, 109
Scire, 110
Scripts, 180, 181
Searle, John R., 193n.20
Self, 13
Self-other relations, 167
Semantic content, 179
Sensus communis, 86, 88
Seriality, 168, 169, 170; dialectic of series and group, 171–72
Shelley, Mary, 51
SHRDLU, 178, 181
Signs, 21
Singer, Charles E., 104
Skepticism: ancient, 33; premodern, 53
Skills, 19, 37, 112
Smith, Adam, 140
Socialism, 51; scientific, 149
Society, debureaucratized, 164
Socrates, 33–38, 40, 44, 66, 67; Enlightenment reconception of, 44
Sophocles, 59n.51
Stapledon, Olaf, 93–95
Studies, science and technology, 16, 102, 105
Subjectivity, individual, 163

Sublime, 80, 81
Summum bonum, 70
Summum malum, 70
Suvin, Darko, 92
Switch, gestalt, 126
Symbols, 21, 176
Syntactic relations, 179
Sypher, Wylie, 58 n.48
System: artificial natural language understanding, 189; epistemically closed, 181; epistemically open, 186, 189; language processing, 180–84
Systems: biological, 189; classifier, 206, 207; expert, 196, 197, 198, 199, 201; logic-based, 196–200

Taylor, Frederick, 143, 155, 159
Taylorism, 149, 159; as the "scientific management" school, 158
Technai, 35, 36, 37, 38
Techné, 14, 21, 36, 38, 44, 109, 111
Techné-logia, 109, 112, 114
Technical engagements, 32, 43
Technics, 32, 36, 40; delimitations of, 48; Socrates on the utilitarian function of, 40
Technique, 97; and creation, 87; as self-augmenting system, 97
Technique, la, 74; and the machine, 74–75; as mentality and method, 74
Techniques, autonomous, 21
Technological fix, 135, 144
Technology, 13–24, 31, 32, 36, 41, 46, 48, 89, 91, 123, 175; and actuality, 91; and administration of work, 157; affirmation of, 40; and alienation, 96; ancient critique of, 33–39; as artifact, 13, 107; as artifice, 107; as an assemblage of concepts, 104; and authenticity, 32; and automation, 149; the autonomy of, 18–19, 117–30, 133–51; benefits of, 45; and bureaucracy, 19, 20, 60–62; and community, 15; and competition, 144; conceptions of, 16; concept of, 13, 23, 107, 114; as control, 15, 47; and the creation of concepts, 102–16; as the creation of intermediary concepts and practices, 18, 103–4, 114; and the critical uneasiness of romanticism, 48–49; definition of, 14, 15, 23, 105–6; de-

velopment of, 121; discourse on, 13, 16, 21; and discourse, 14, 111; dissemination of, 14; engagements with, 13–24; as equipment, 60–71; the equipment of, 23; etymological and lexical deliberations on, 104–13; examples of, 124; and the future, 112; and God, 46; and the good life, 66; as habitat, 13, 107; as the hand-maiden of science, 117; as a Heideggerian epoch of Being, 108; heterogeneity of, 23; history(ies) of, 105, 145; humanity and, 17, 31, 33, 41; and human nature, 20; identity of, 13, 14; and the independence of human action, 120; the inherent goodness of, 39; as integral part of science, 123; the issue of, 103; joint articulation of instrument and concept, 107; and knowledge, 45; and labor-saving devices, 98, 132–51; and the life of contemplation, 71; the life of fabrication, 17; and the life reason, 17; links with scientific theories, 122; logic of efficiency and, 162; and the machine, 96, 98; as machine, 118–20; and material freedom, 53; and mechanistic philosophy, 46–47; mediacy of, 13, 14, 15, 16, 21, 103, 107, 113, 175; as medium, 13; modern, 60, 61; modern, and liberation, 61–62; as a myopia, 89; and nature and artifice, 45, 46–47; the operations of, 113; philosophy of, 105; politics and, 35–36; politics of, 105; and possibility, 91; presentation of, 114; and problem solving, 76–95; as a process, 121; the process of, 119; as product of science, 122; the question of, 13–24, 67; as reified, 117, 130; as a relay, 108; and Renaissance and Enlightenment optimism, 39–46; representations of, 14; and responsibility, 132–51; and romantic ambiguity, 33, 46, 46–55; and a sacred order, 99; and science fiction, 76; and scientific knowledge, 53; and the search for absolute methodology, 95; as self-creative act, 53; as self-dismantling process, 108; as self-supplementation, 104; the sensible presence of, 14; as skill, 60–

71; the social context of, 128; sociologies of, 105; as something concealed and mysterious, 102, 103; and the sublime, 53; as tools, 118, 120, 122, 129–30; as transfiguration and transformation, 109–11; and transformation of the labor process, 143; and translation, 13, 24, 104, 106, 109; as types of knowledge, 60–71; ubiquity of, 13, 14, 15, 16, 21, 103, 107; the usefulness of, 112; willed for humanity by God or Nature, 45

Tektonikos, 34

Telescope, 124, 125, 129

Thales, 66

Theobald, Robert, 148

Theory, 123, 128; critical, 161, 163, 164; critical, and the logic of rationality, 162; mainstream organizational, 163; organization, 160; systems, 163

Tools, 176

Topoi, 95

Topology, four-color theorem of, 201

Tradition, authoritarian, 134–40, 135

Translation, 109, 111; as creation of intermediary concepts and practices, 109; as displacement, 109

Tupperware, 37

Turing, Alan, 196

Understanding, 13, 23, 178, 182; natural language, 175, 177, 186, 190; natural language processing, 181

United Automobile Workers, 144

Universali fantastici, 82

Universali intelligibili, 82

Unknown: scientific, 74; technological, 74

Ure, Andrew, 139, 141, 143, 146

Usage, artifice of, 109–10

Use, 109, 121, 129, 130

Useful, the, 118

Utopia, 95, 96, 138, 148

Utterback, James W., 24n.13

Values, 15

Veblen, Thorstein, 152n.17

Verene, Donald Phillip, 85

Versuch, 103

Verum factum, 85

Vico, Giambattista, 18, 74–101

Virtue, 49, 64

Voegelin, Eric, 69

War, nuclear, 89

Weber, Max, 20, 135, 136, 139, 155, 156–61, 163, 172

Weizenbaum, Joseph, 176

Wesley, John, 39

Whitehead, Alfred North, 66, 67, 91

Whitman, Walt, 80

Wiener, Norbert, 39

Will, free, 132

Winner, Langdon, 19, 120

Winograd, Terry, 178, 181, 182

Wiser, James L., 17, 19; on three levels of technological consciousness, 17

Wittgenstein, Ludwig, 14, 105

Wordsworth, William, 47–50, 52

Work: and displacement of workers, 143; and the introduction of new technologies, 134; philosophy of, 136; and workers, 133–51

Workers, 156–62, 165. *See also* United Autobile Workers

World, blocks, 178, 182

Xenophon, 33, 34, 35